MALKY MACHINE COLOURING BOOK

Jack Kirk

For Adults & Children

Copyright 2015 by the Author

ISBN-13: 978-1519722997
ISBN-10: 1519722990

I have a great belief in what the customer wants, the customer gets, and in this case several people expressed the desire for a colouring book. In this day and age, adult colouring books are apparently the in thing. So here it is, The Malevolent Machinery book simply turned into a colouring book. Almost all print removed except this introduction so minds are not strained!

Jack Kirk Dec 2015
jeanjack123@hotmail.com

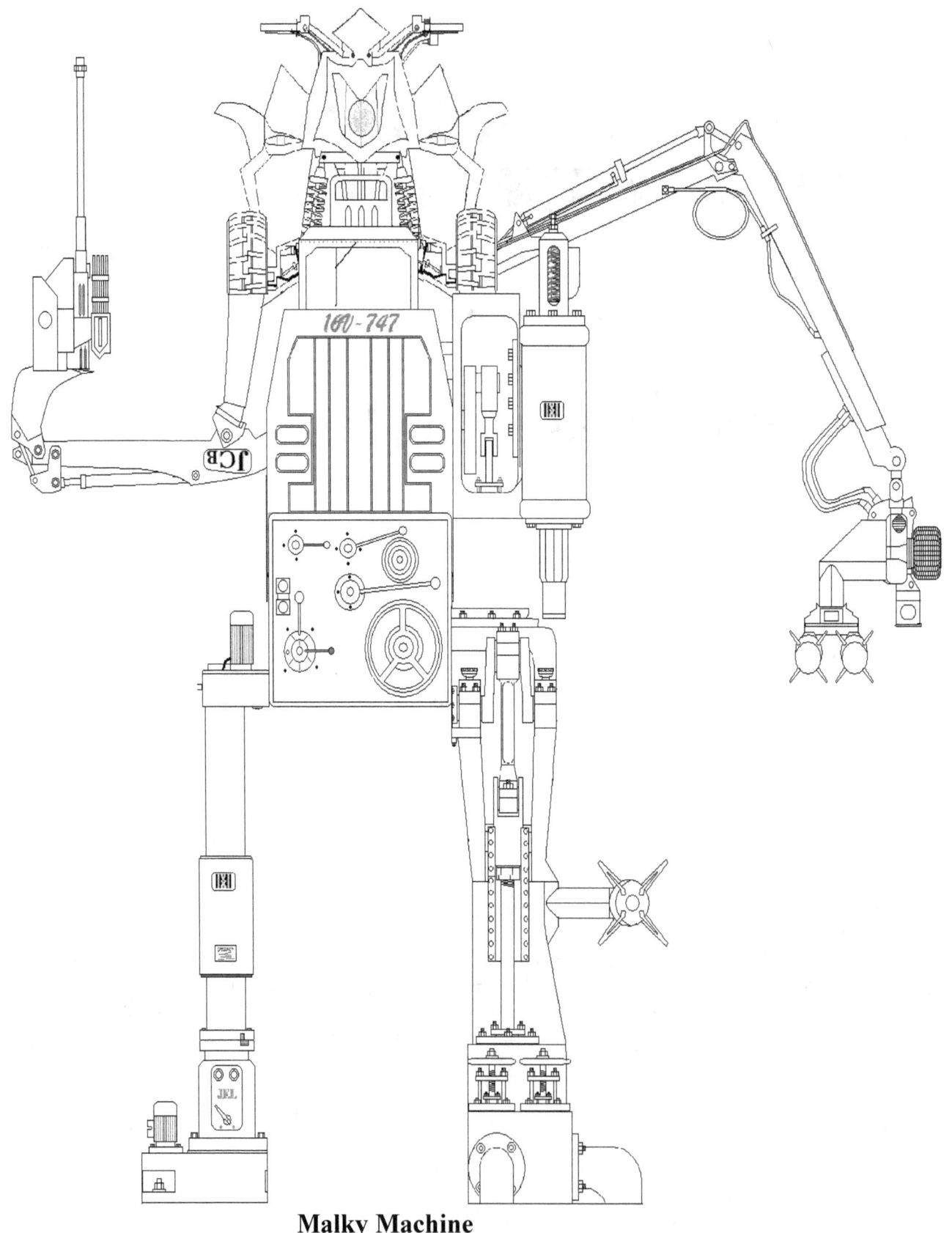

Malky Machine

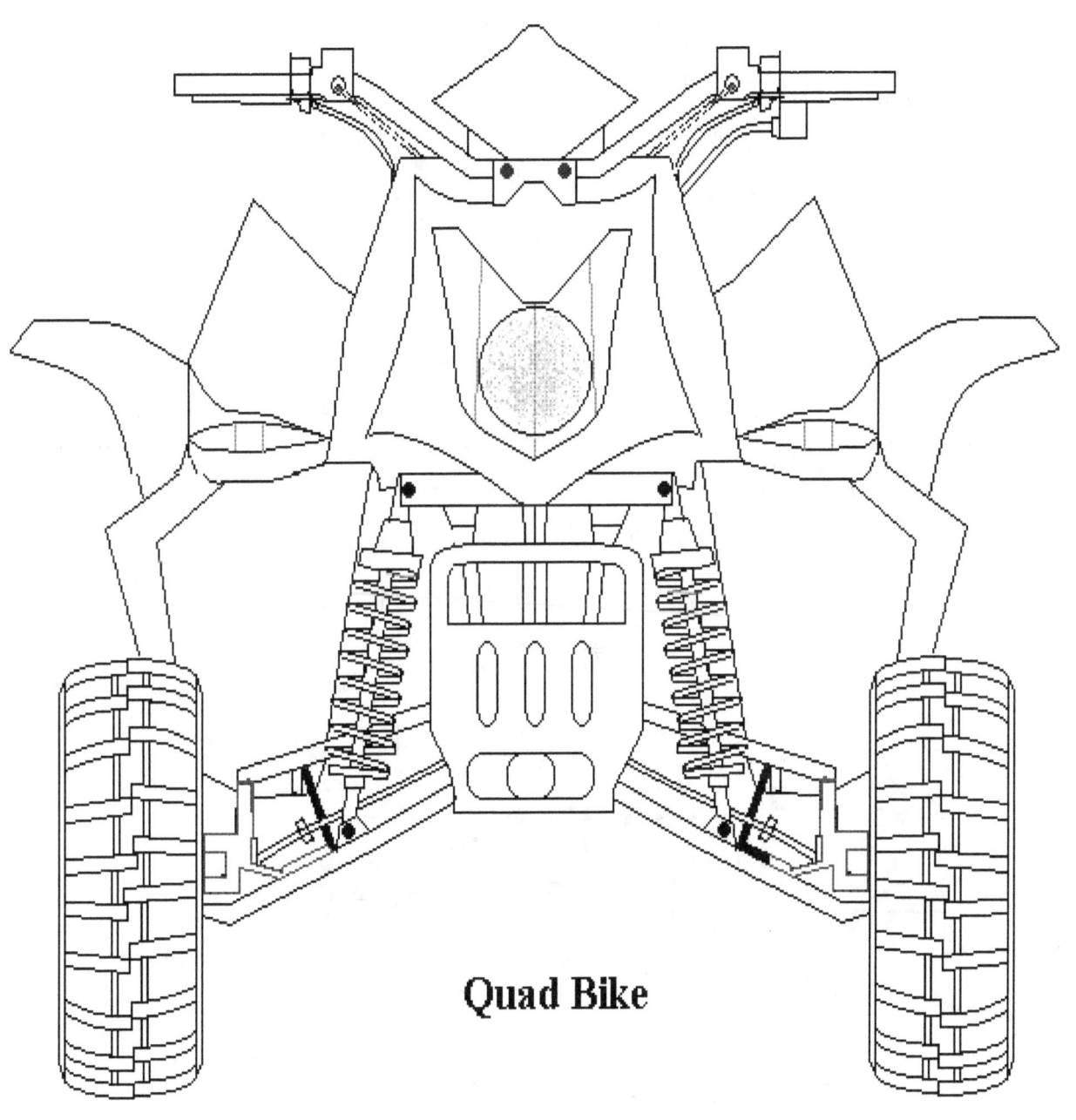

Quad Bike

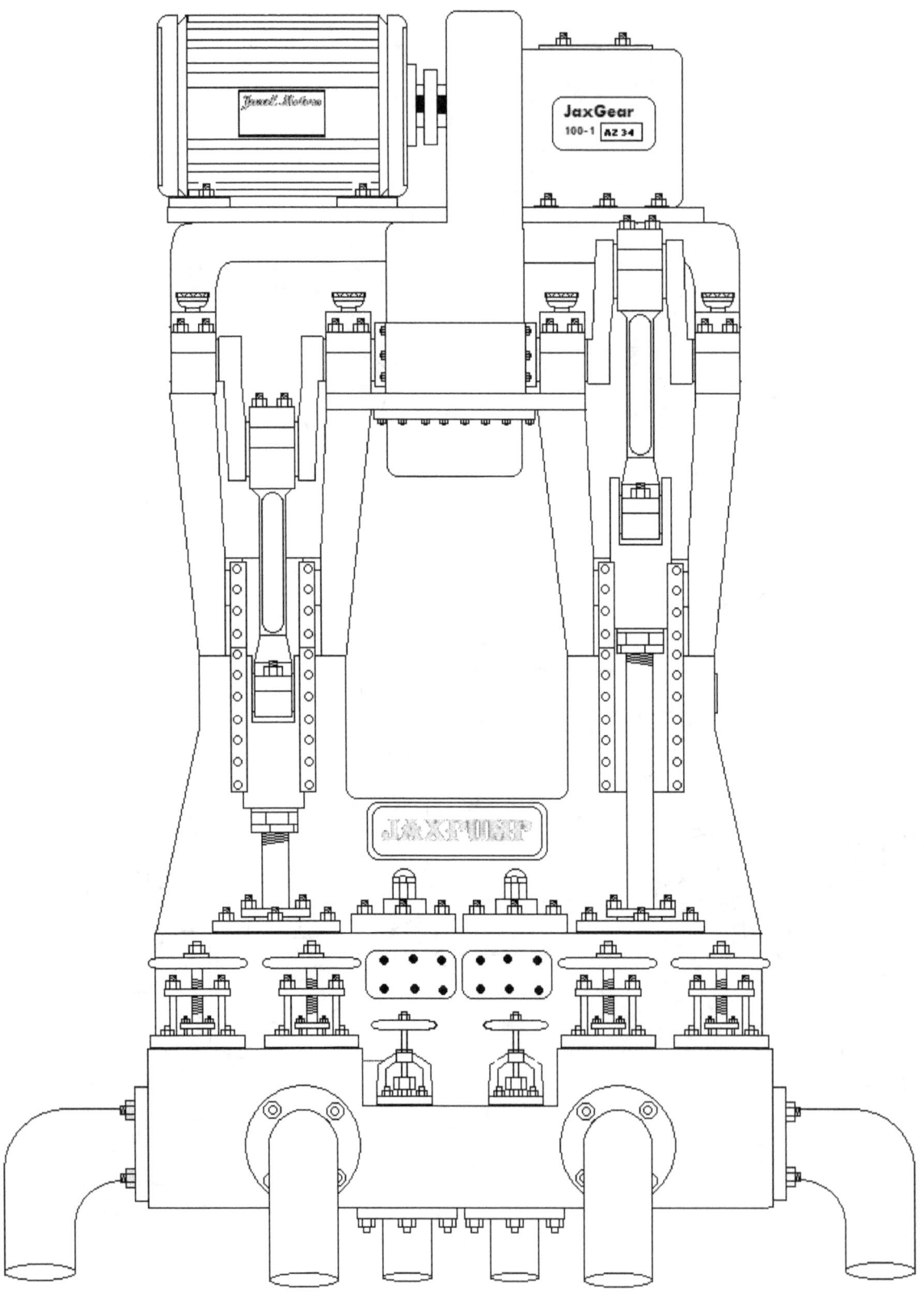

Old Fashioned Reciprocating Bilge Pump

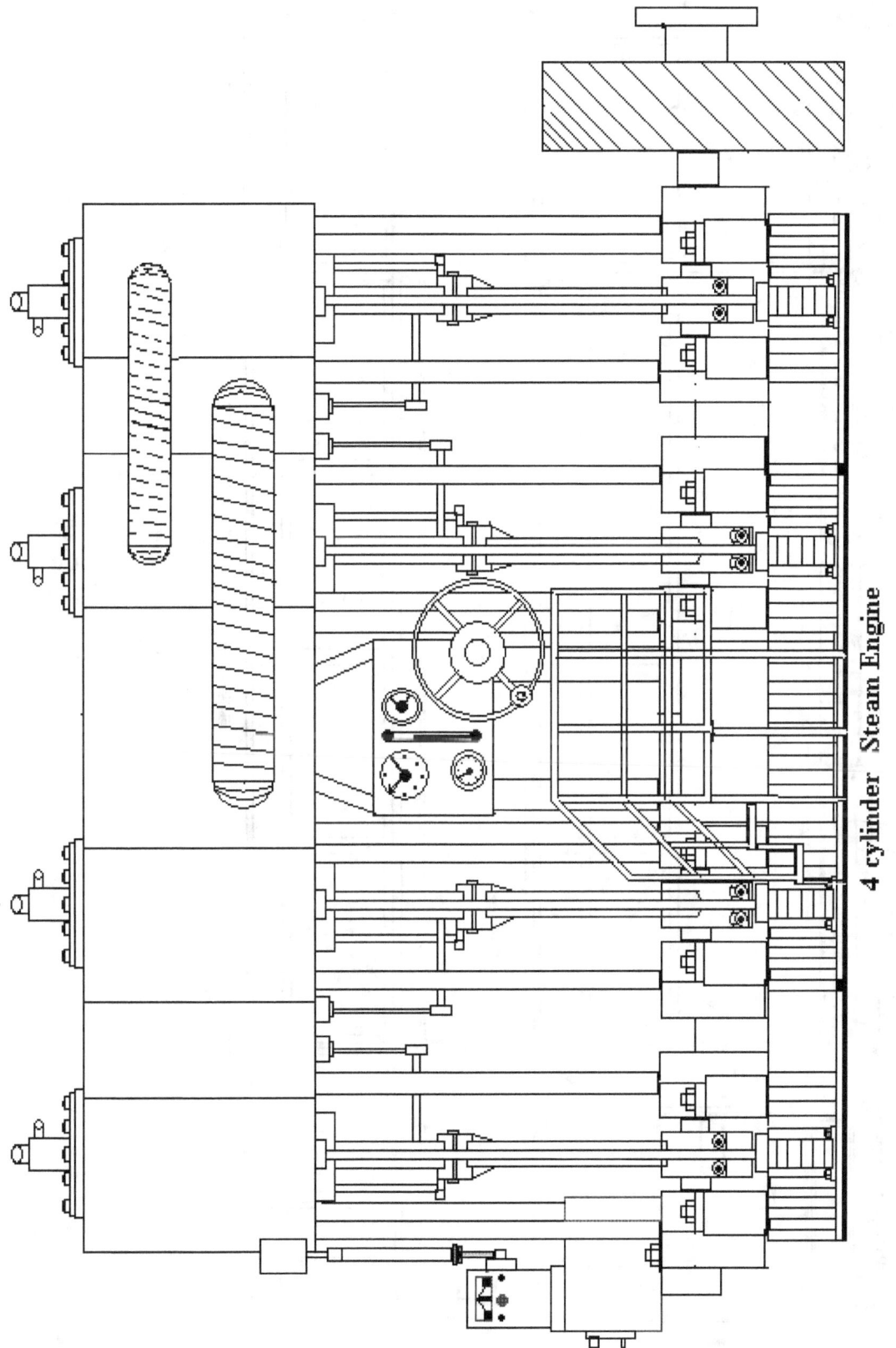

4 cylinder Steam Engine

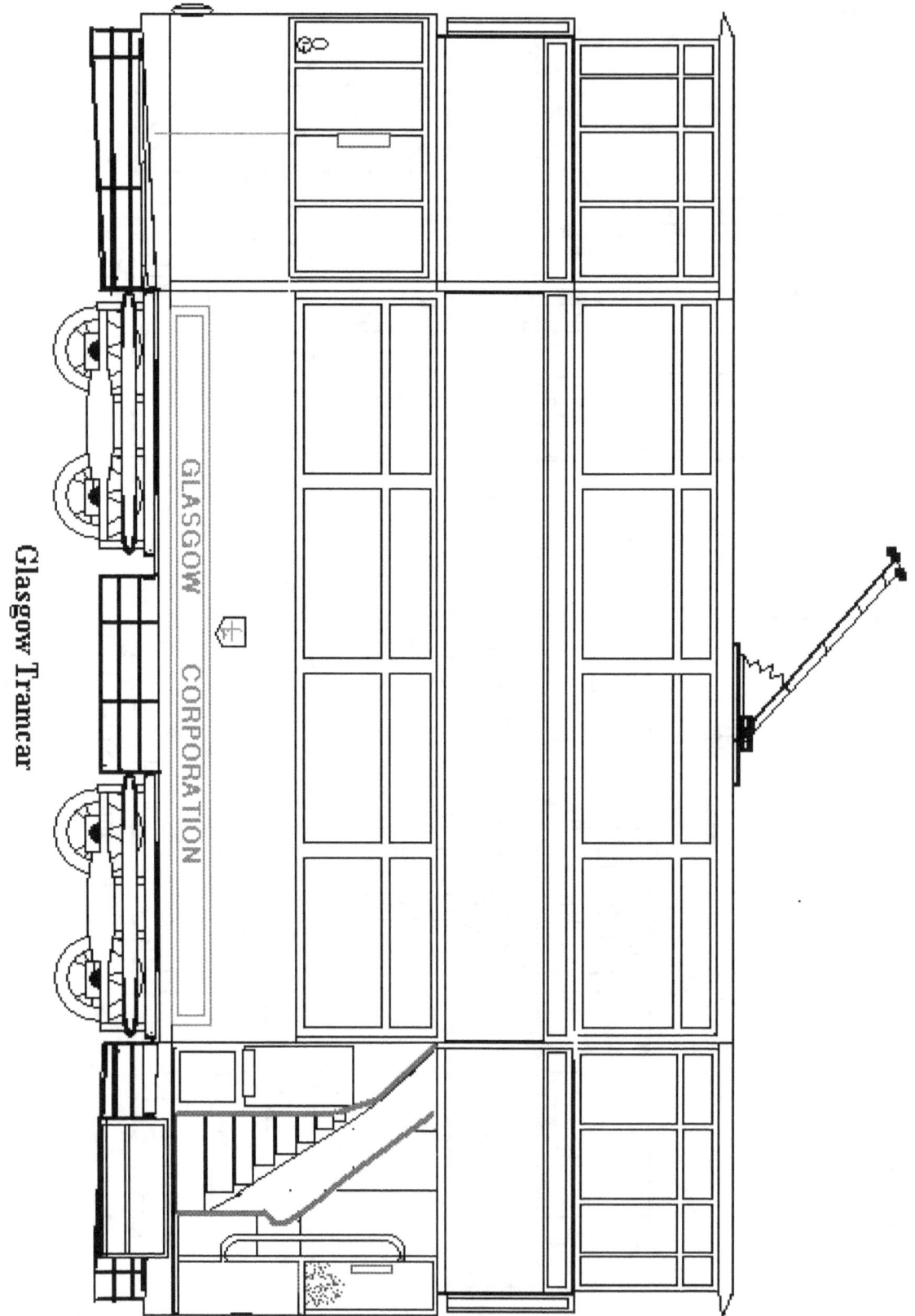

Glasgow Tramcar

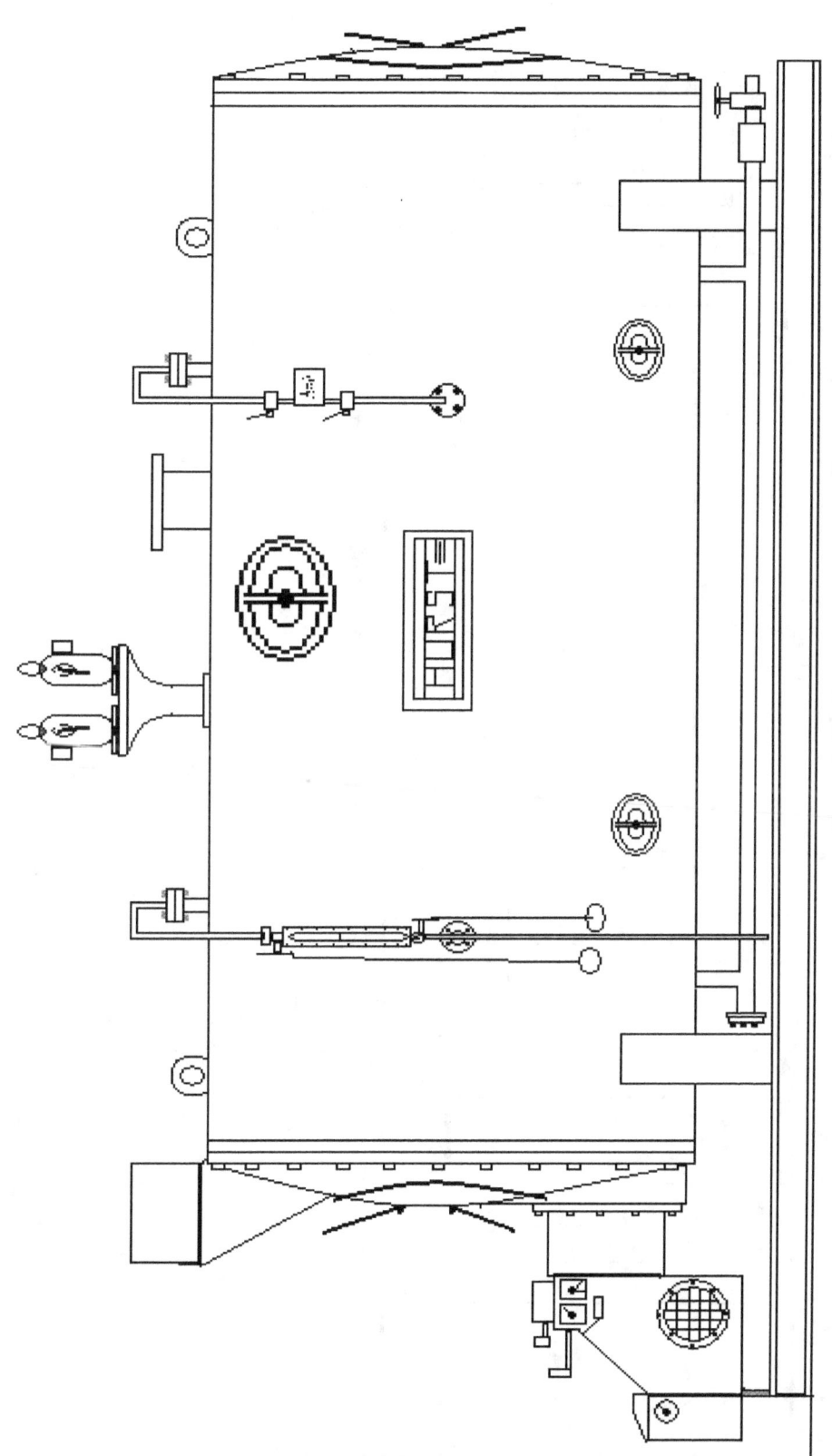

Scotch Boiler

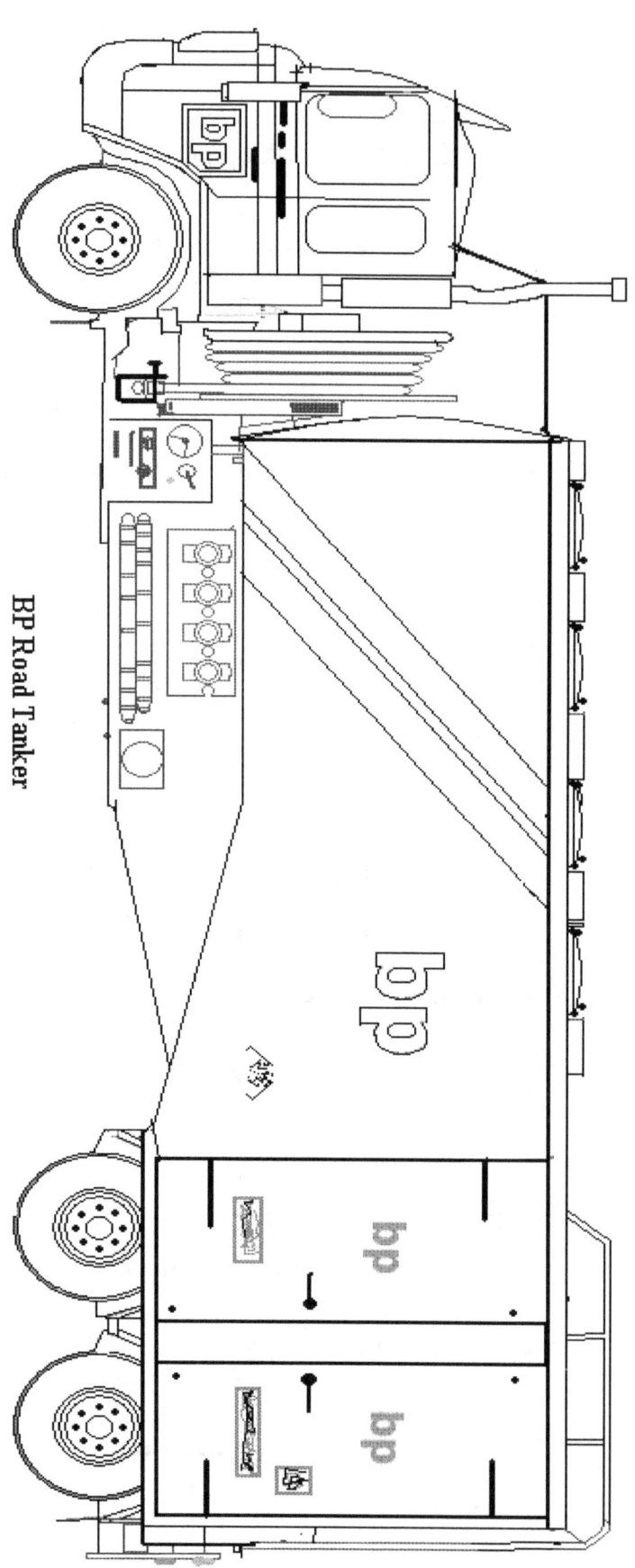

BP Road Tanker

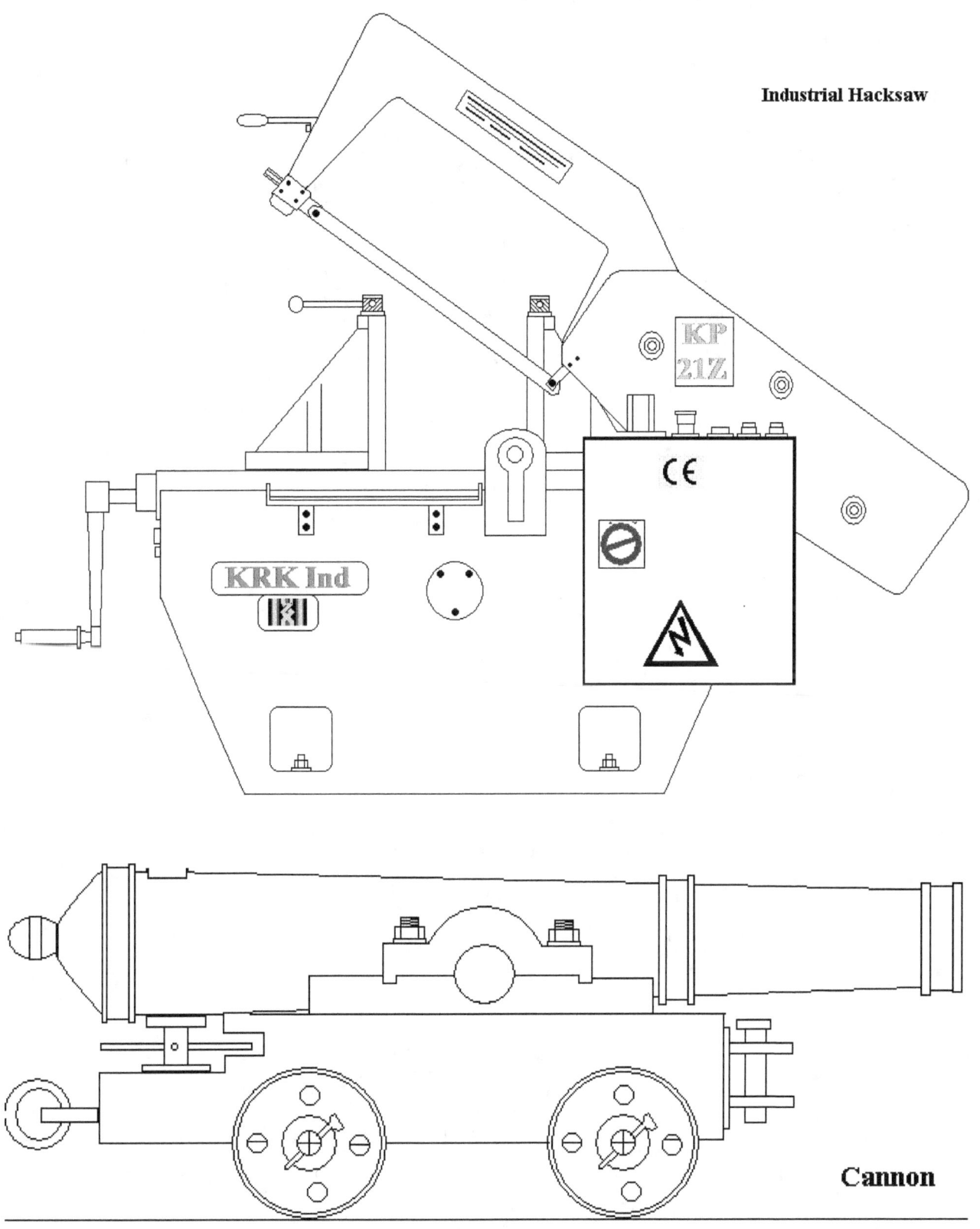

Industrial Hacksaw

KP
21Z

CE

KRK Ind

Cannon

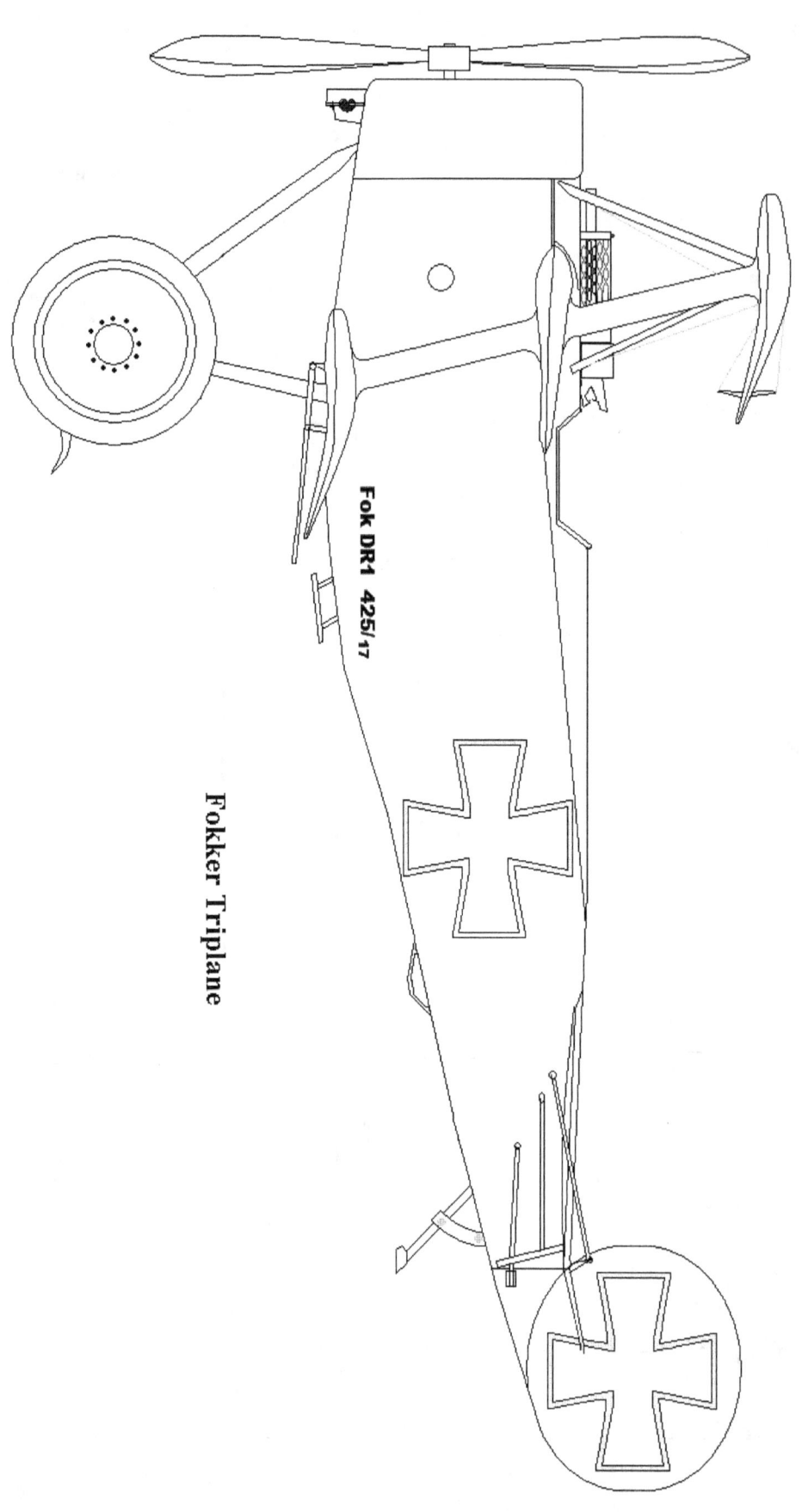

Fok DR1 425/17

Fokker Triplane

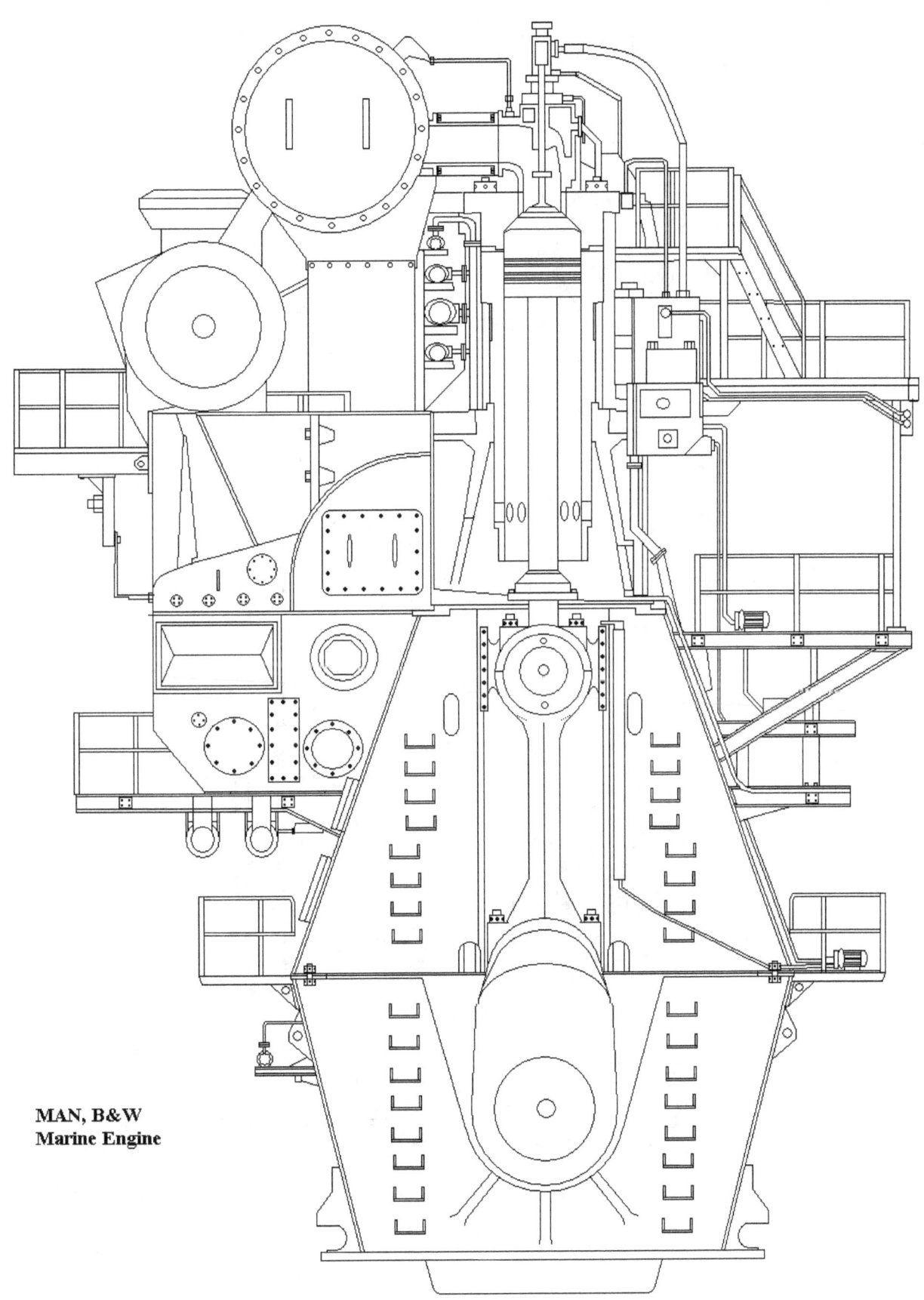

**MAN, B&W
Marine Engine**

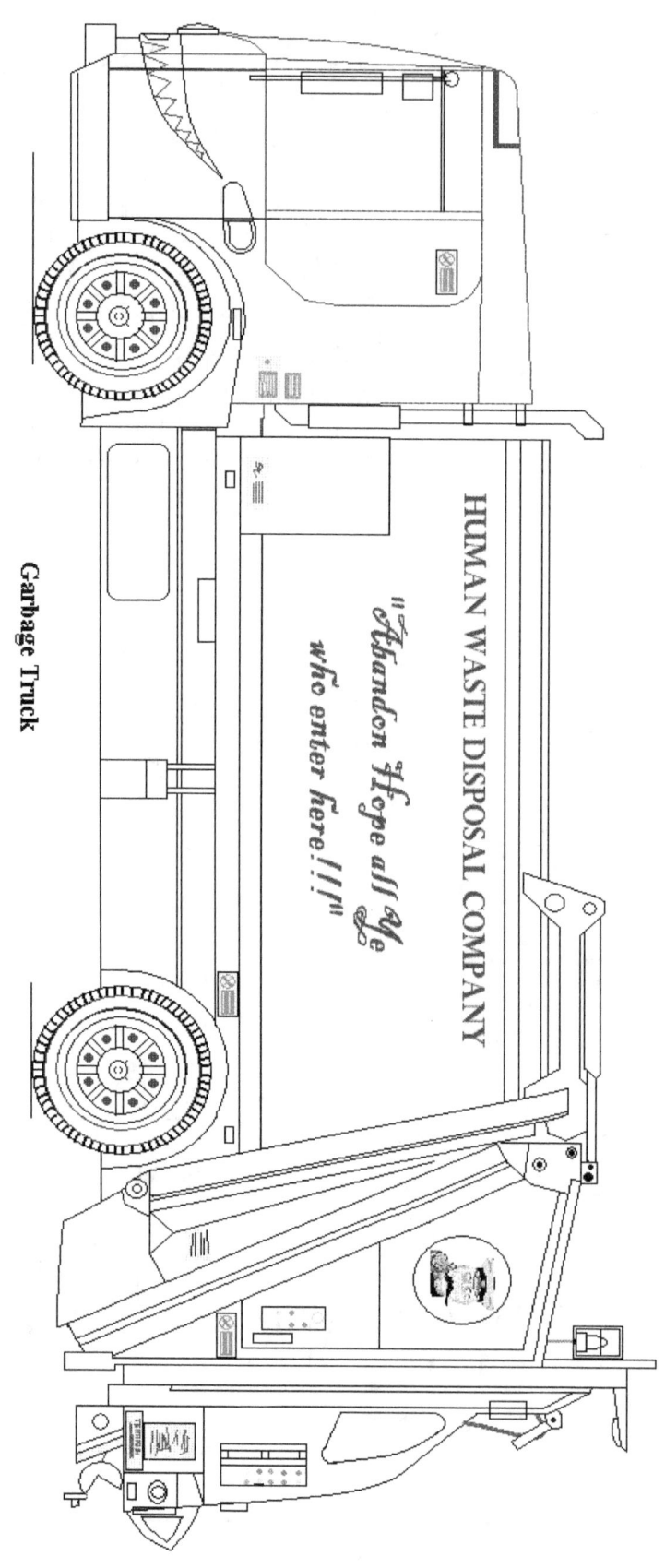

Garbage Truck

Radial Drilling
Machine

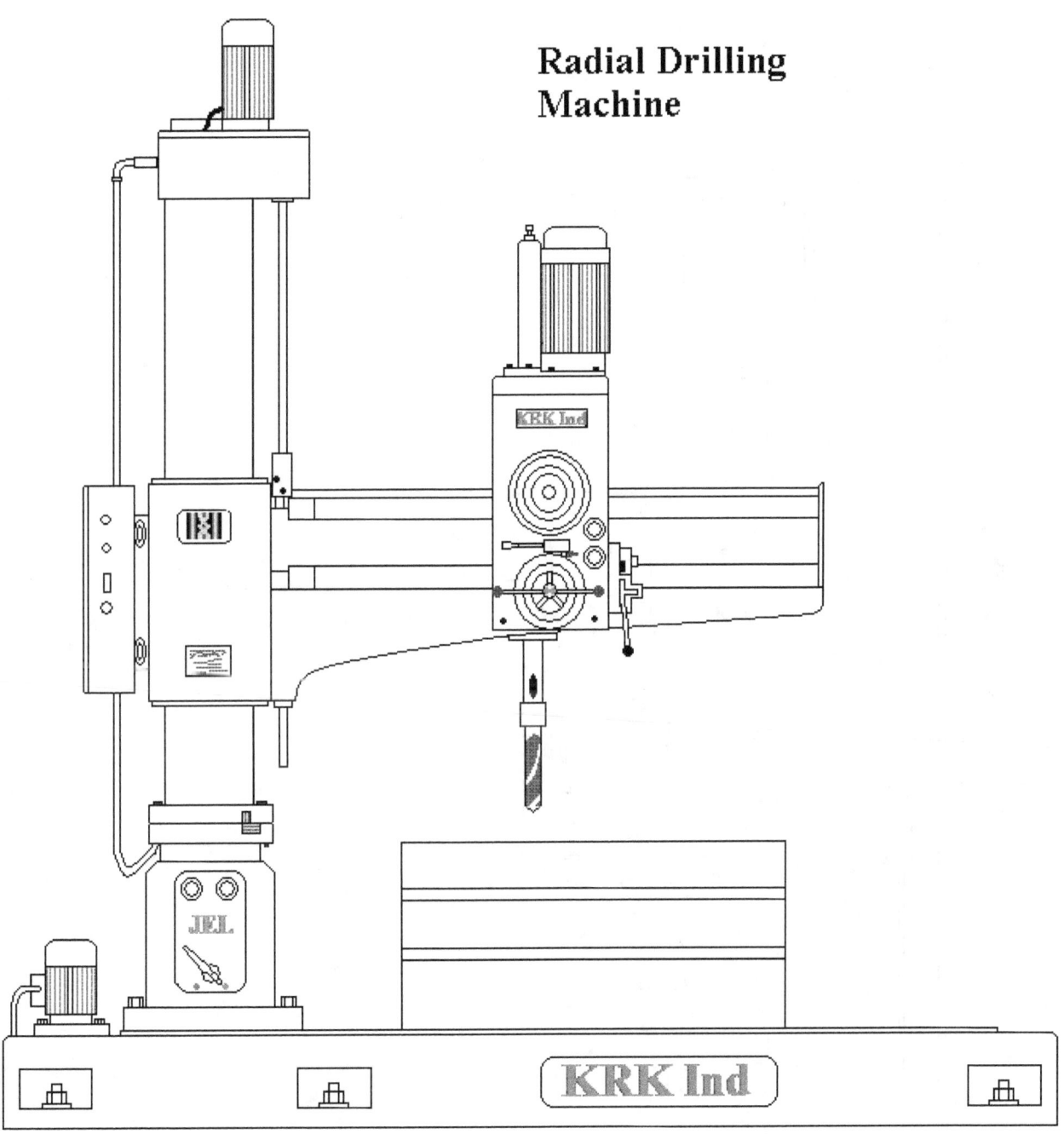

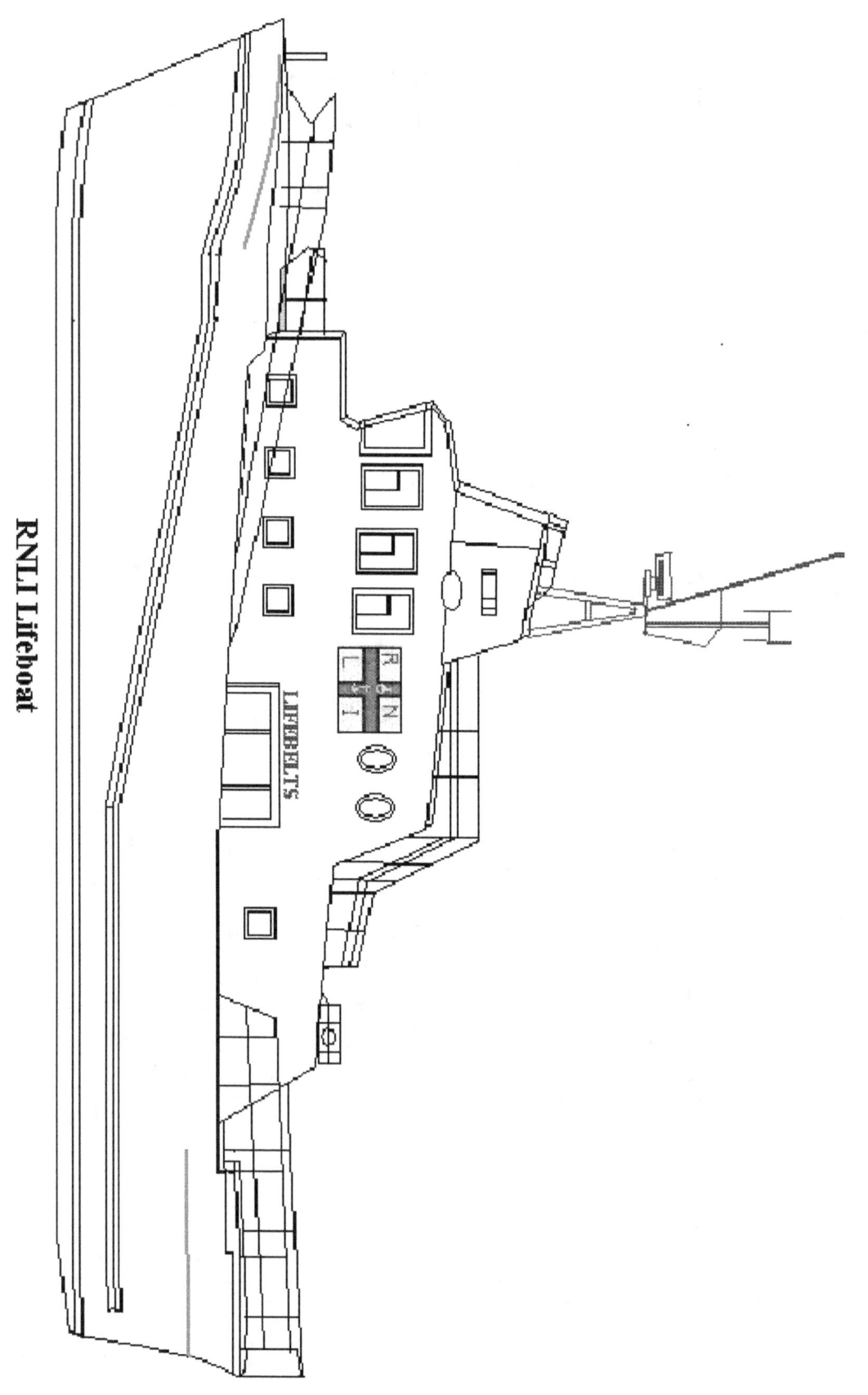

RNLI Lifeboat

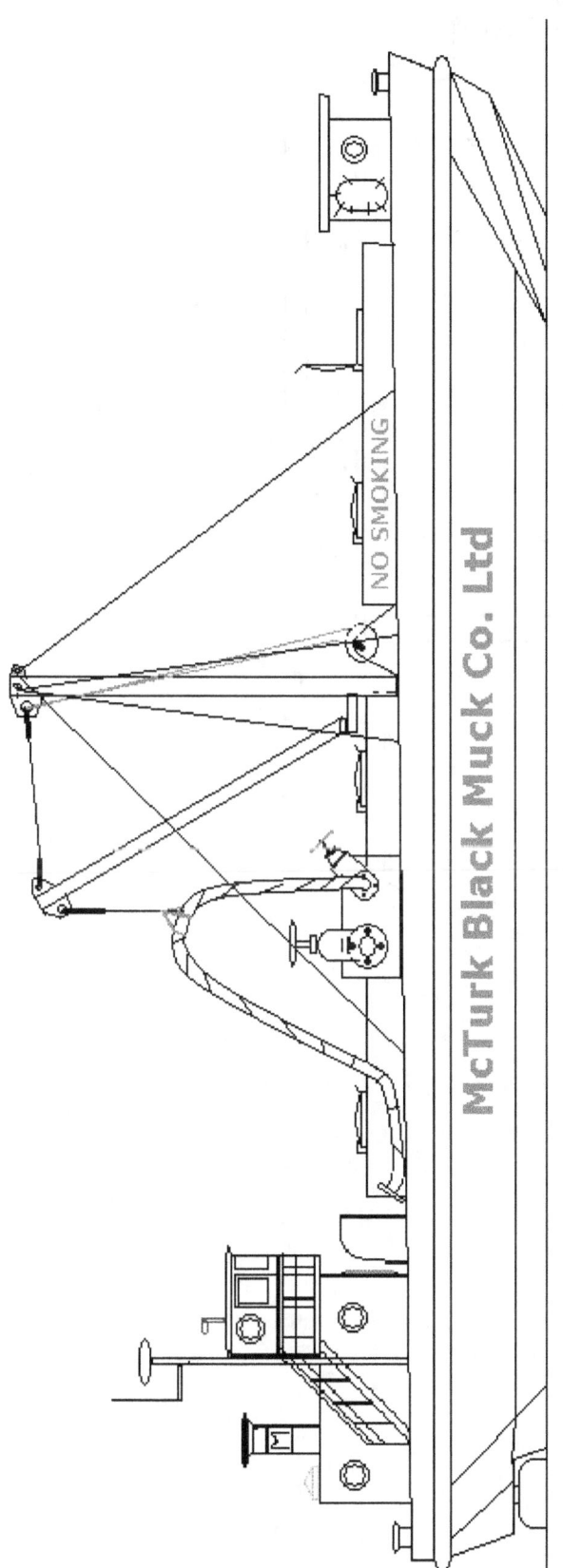

McTurk Black Muck Co. Ltd

NO SMOKING

Ship Bunker Barge

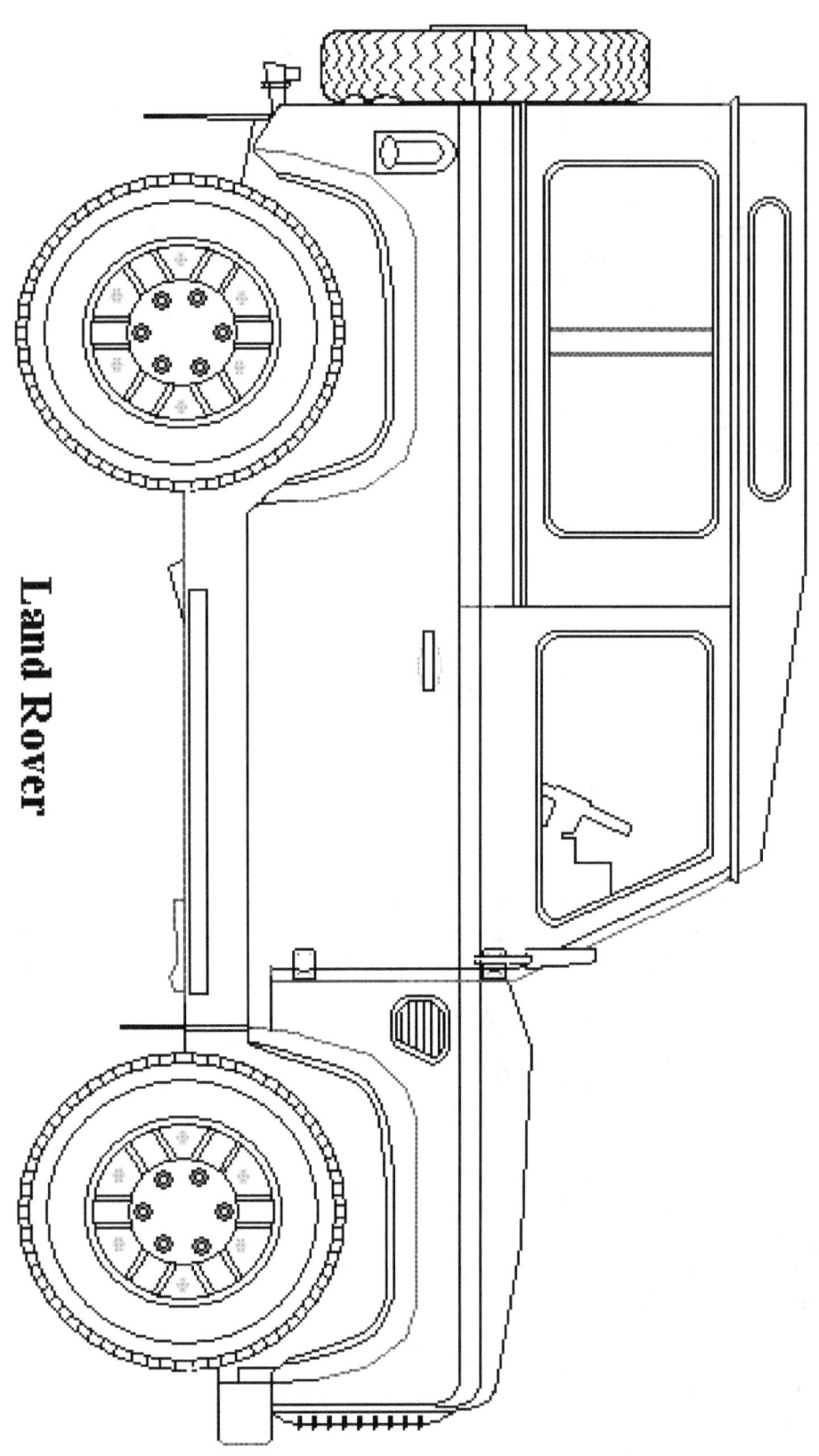

Land Rover

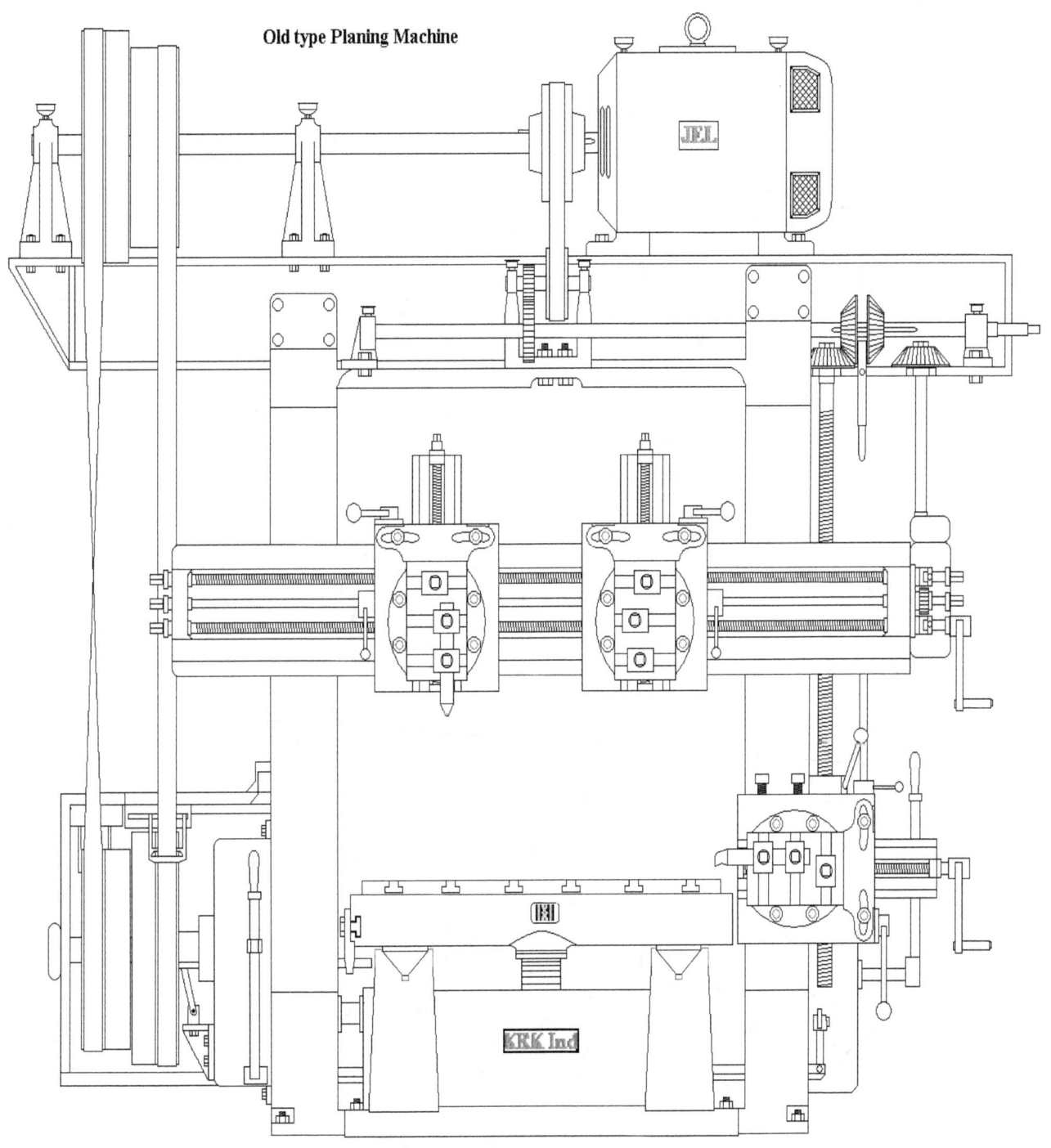

Old type Planing Machine

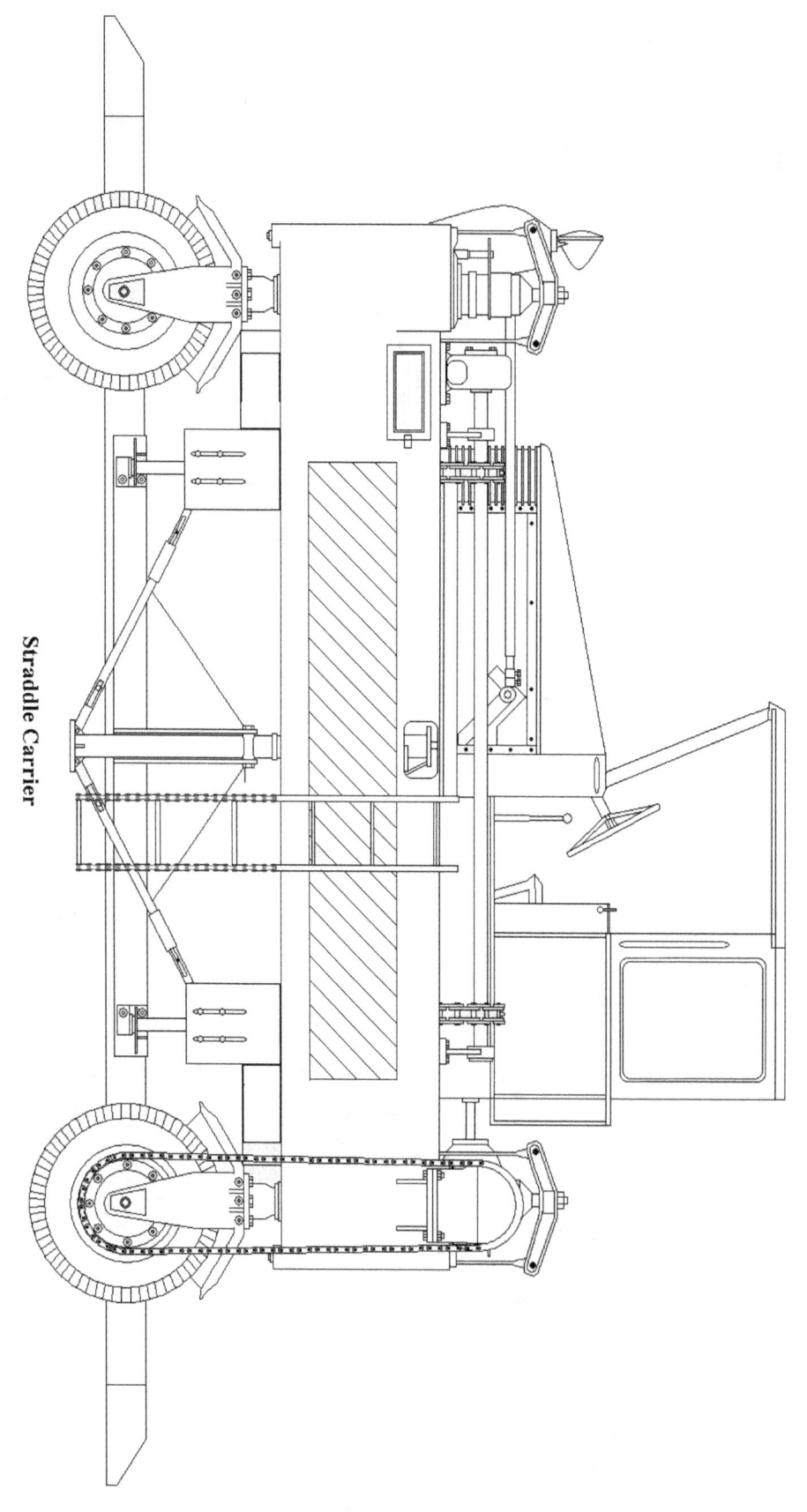

Straddle Carrier

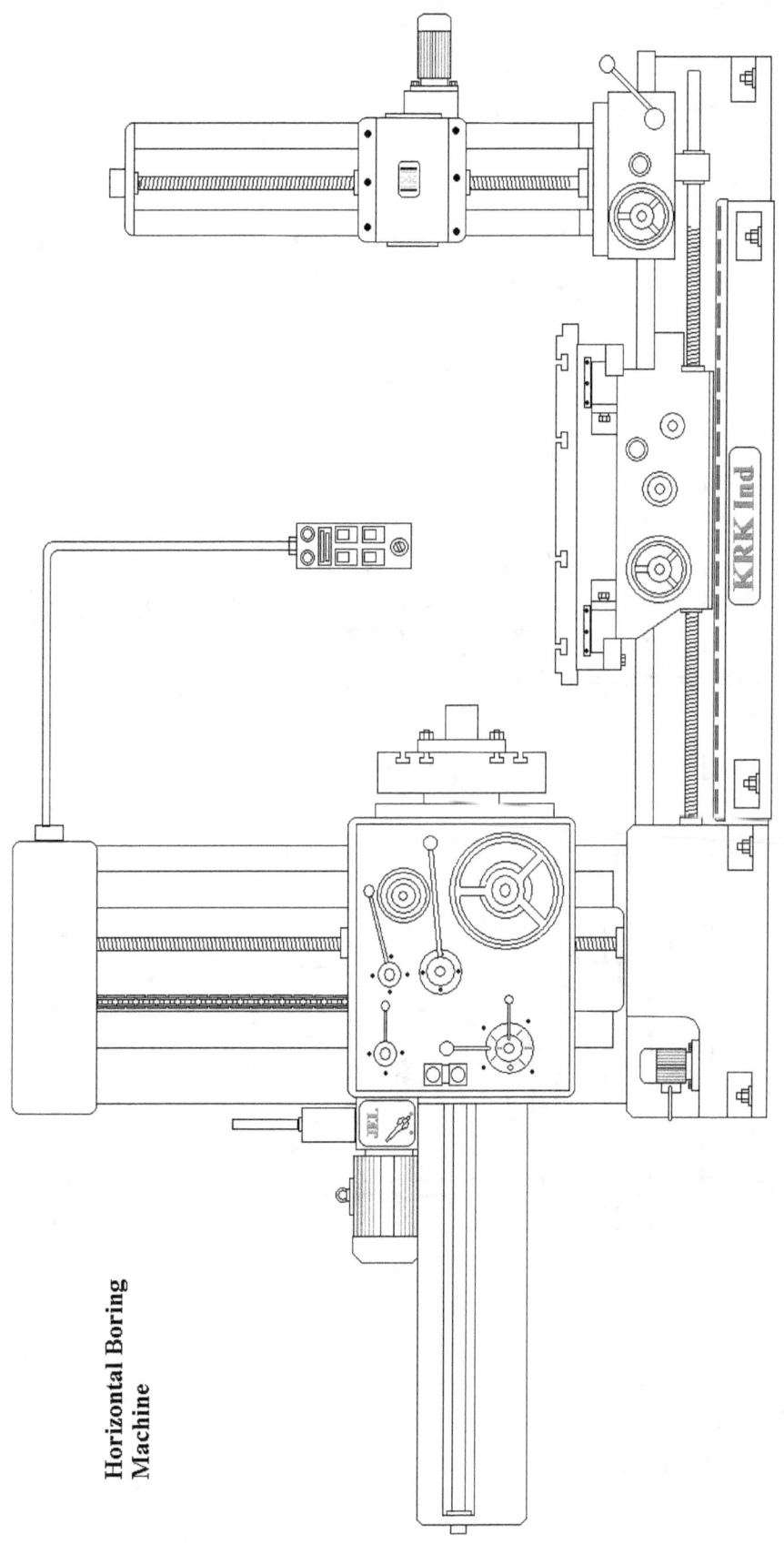

Horizontal Boring Machine

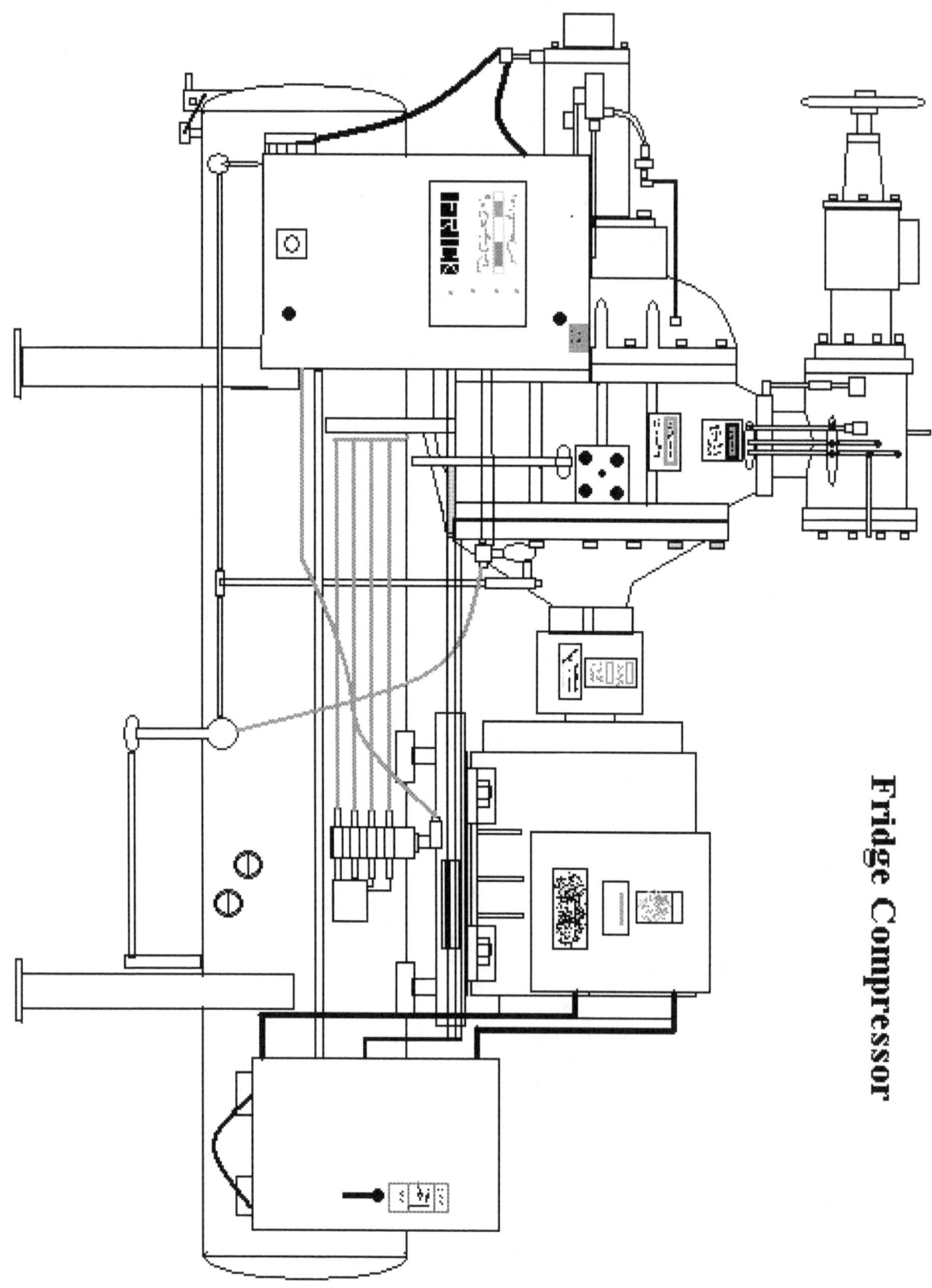

Fridge Compressor

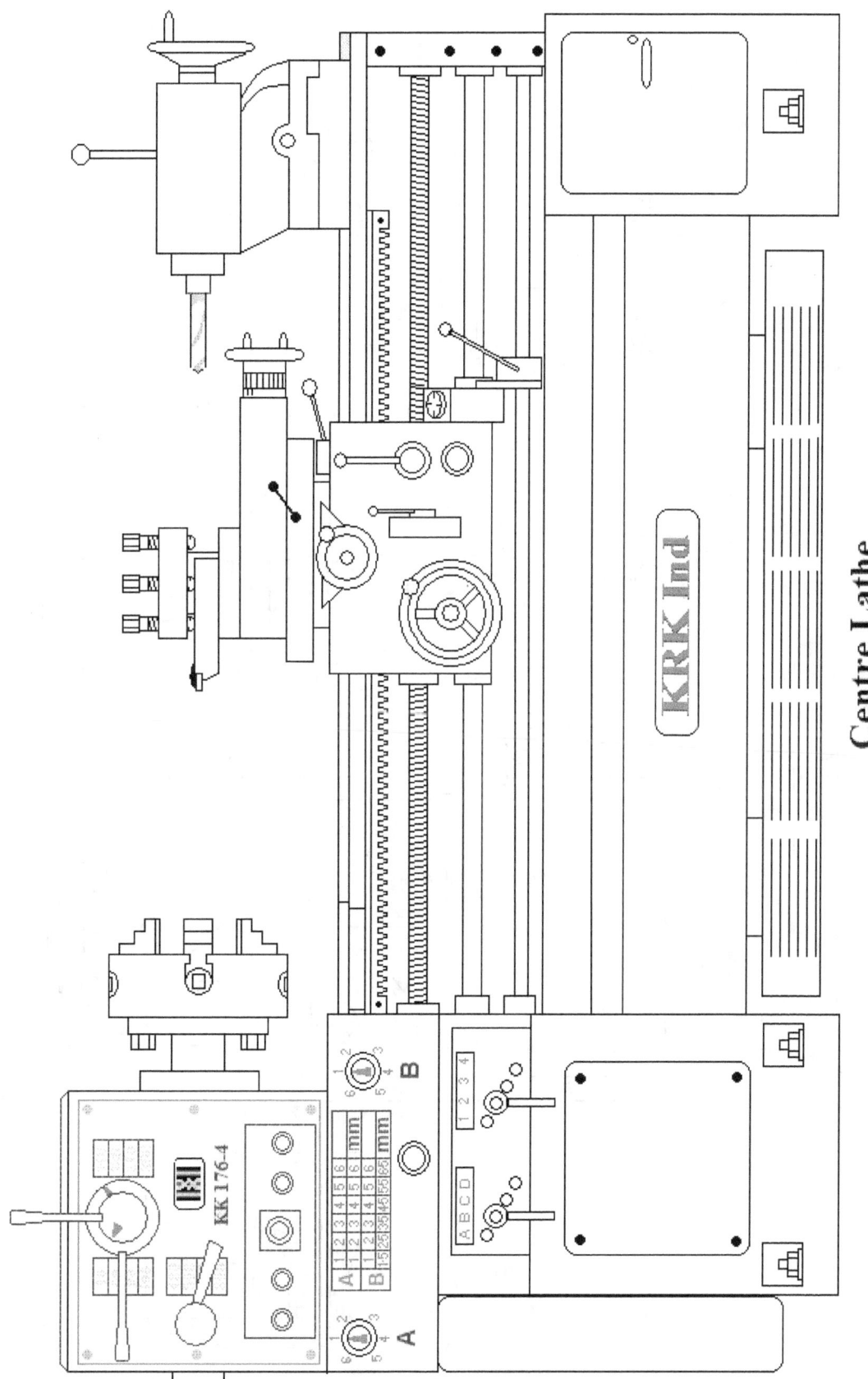

Centre Lathe

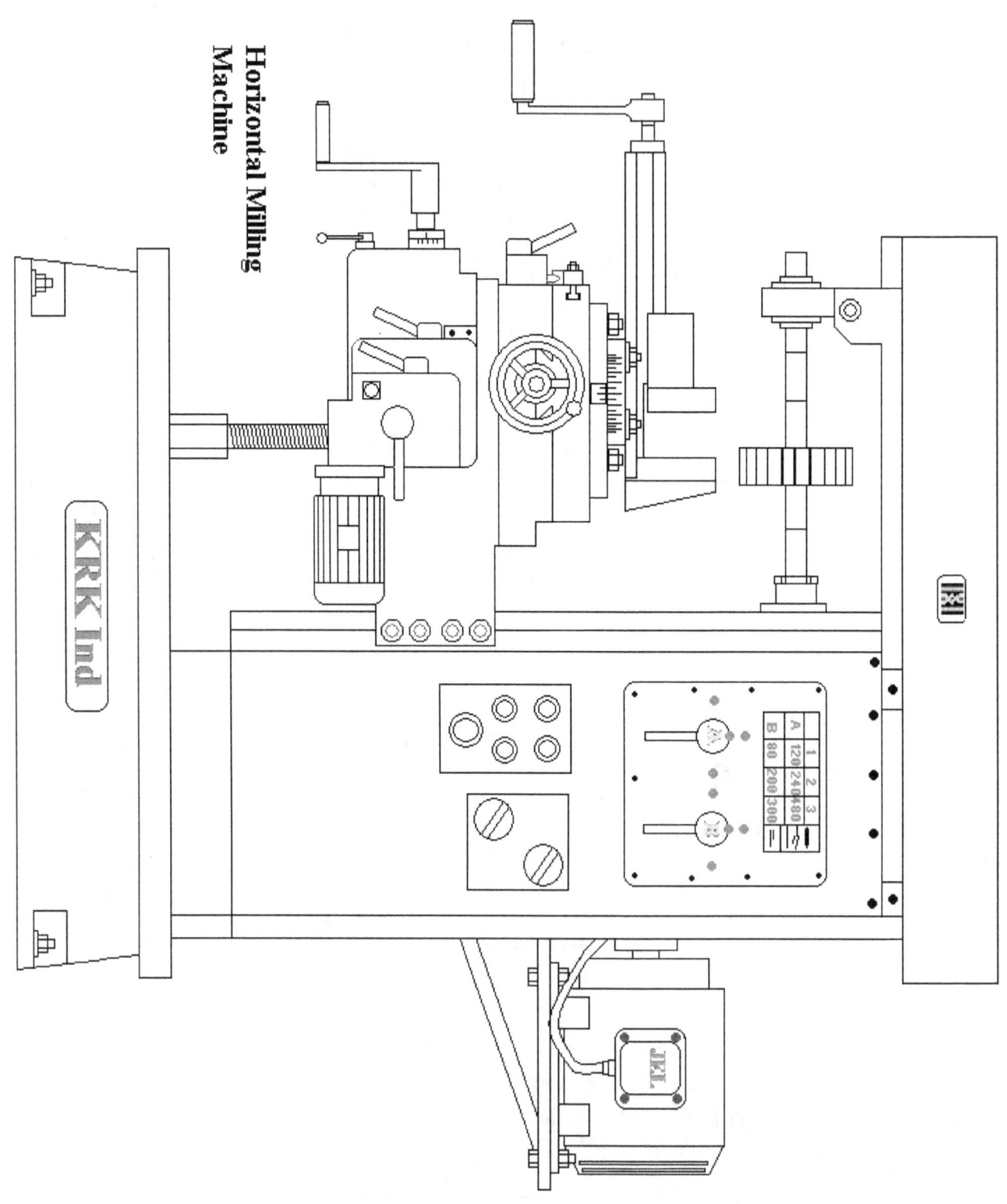

Horizontal Milling
Machine

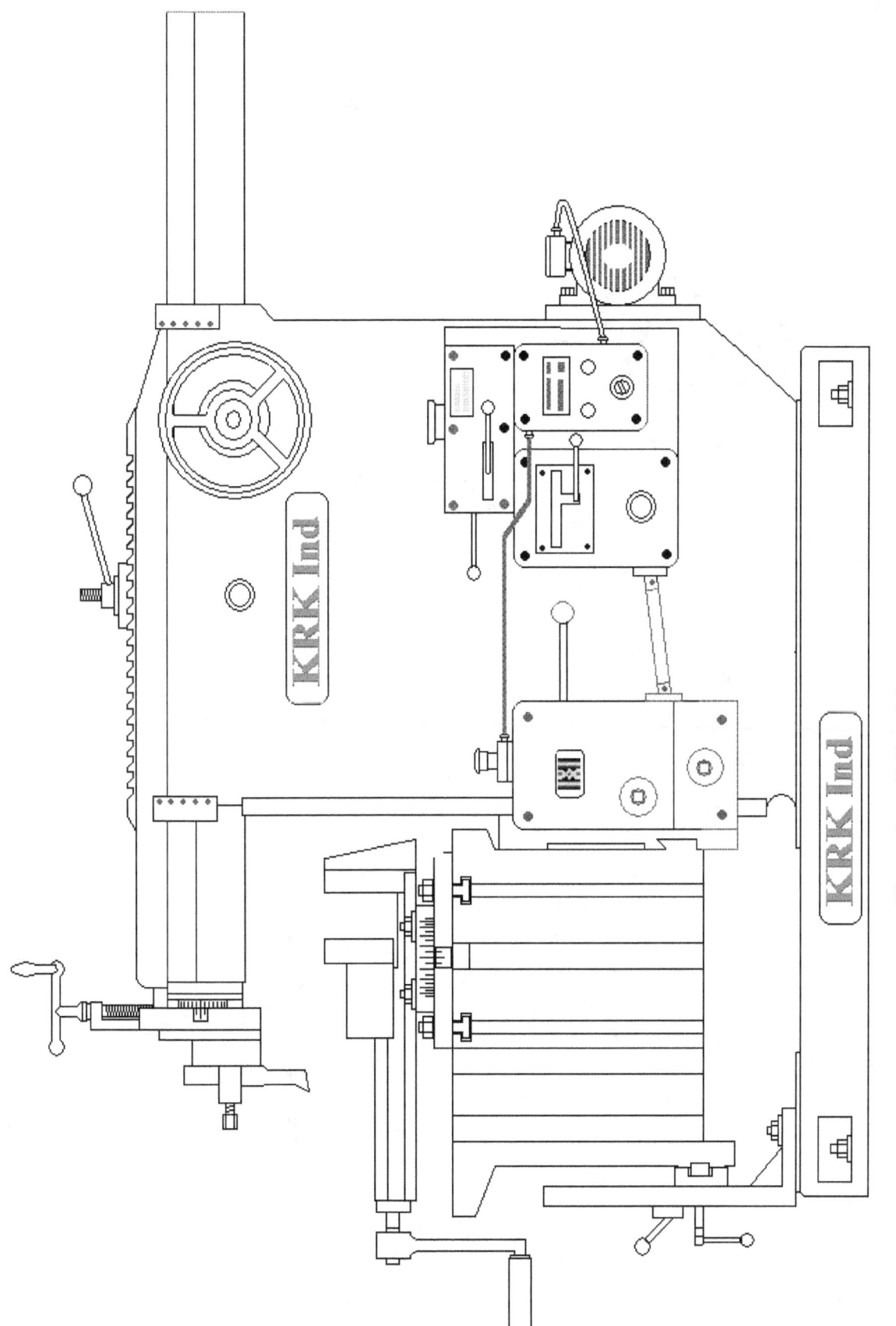

Shaping Machine

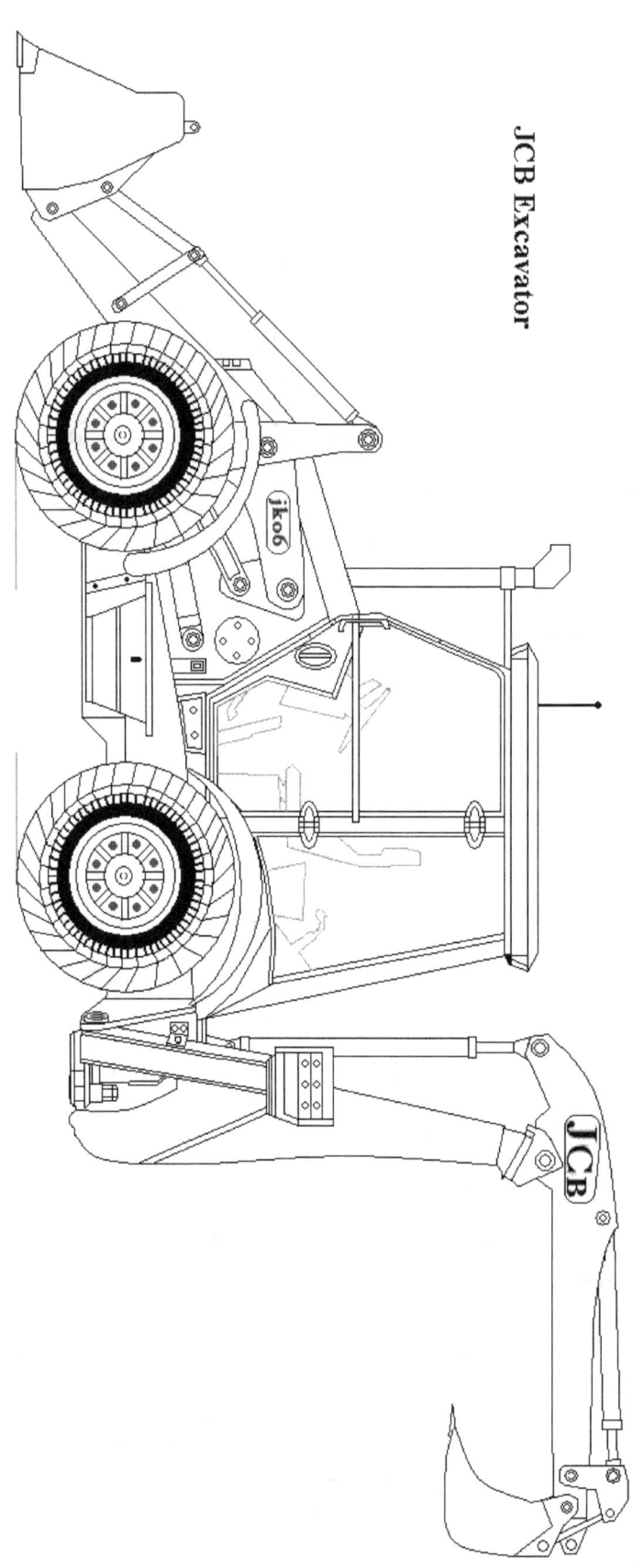

JCB Excavator

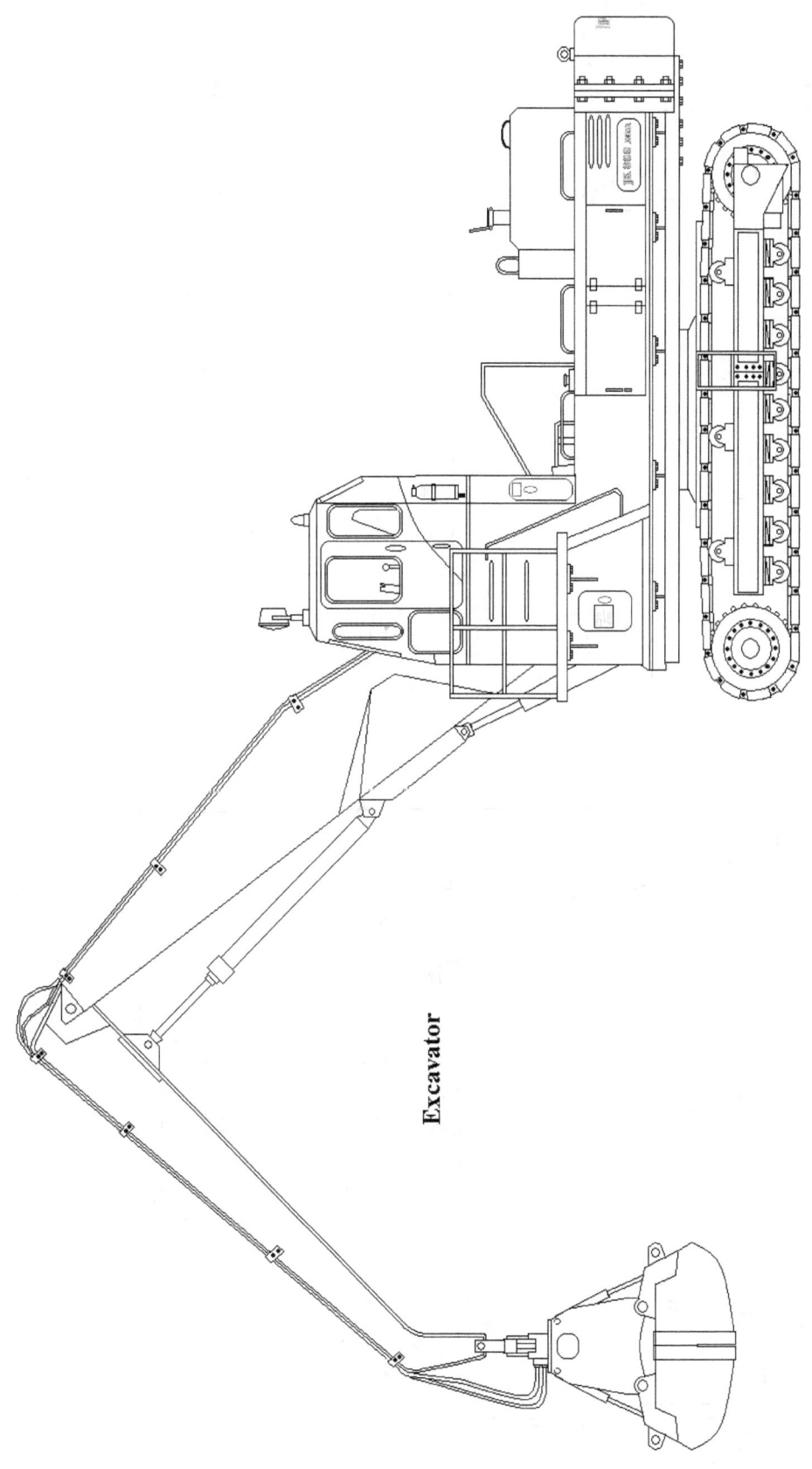

Excavator

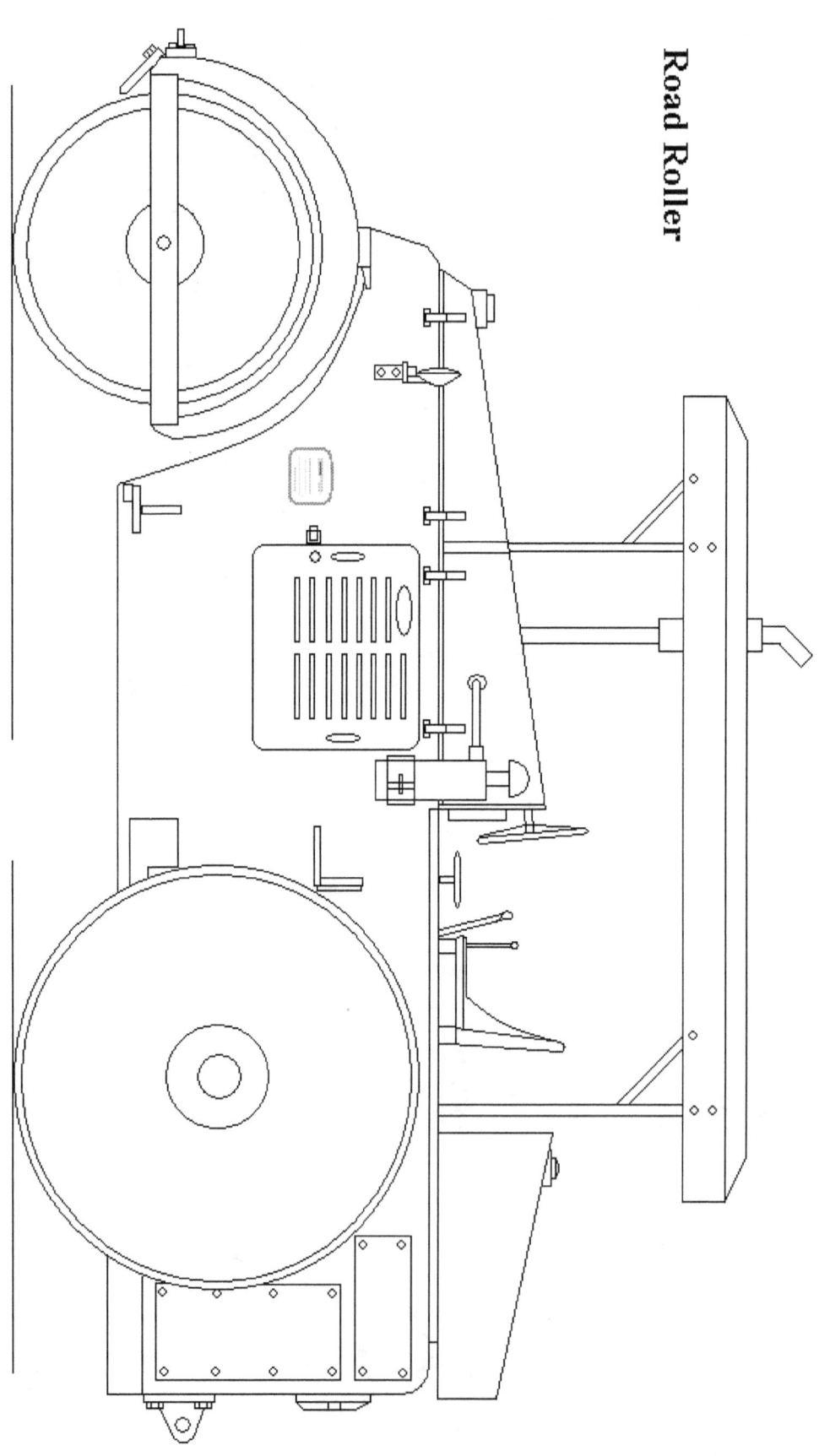

Road Roller

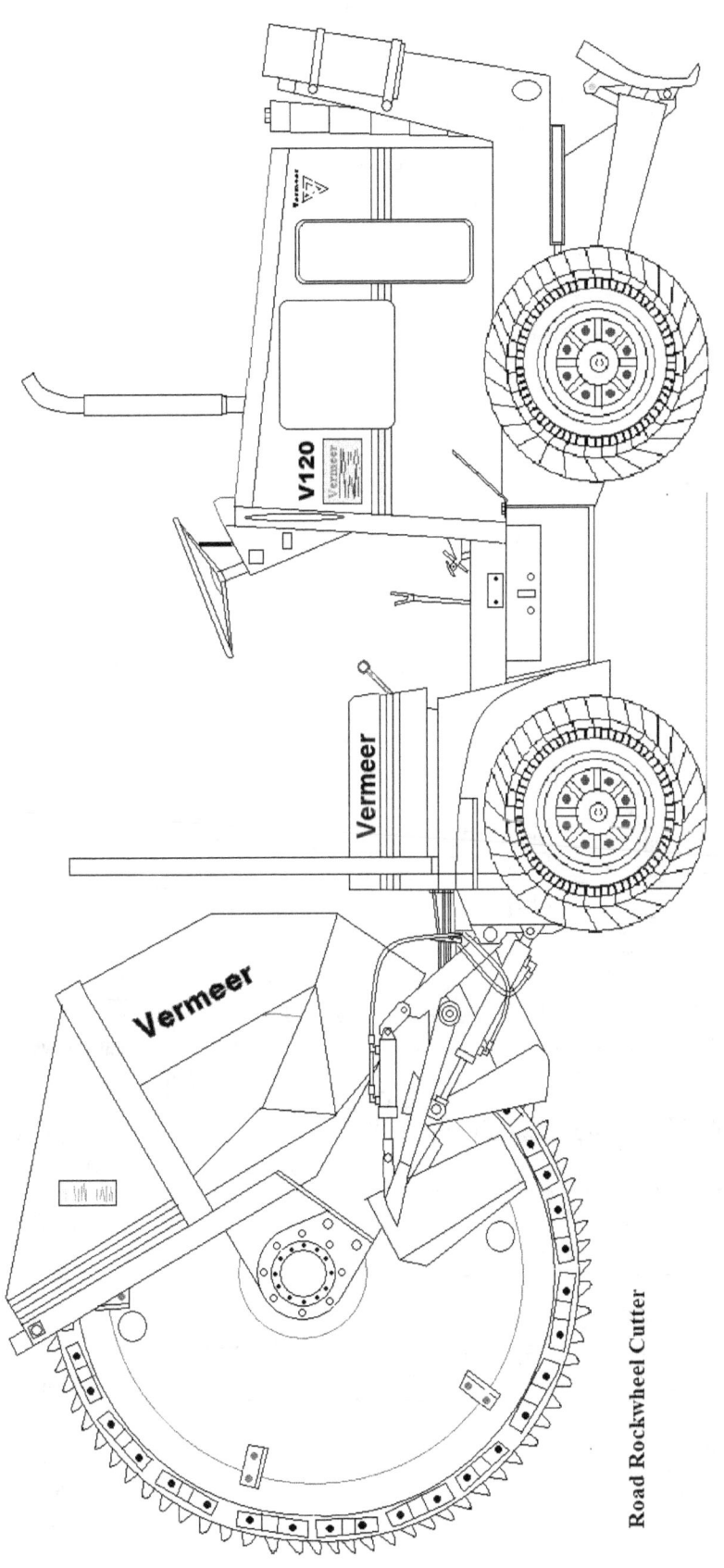

Road Rockwheel Cutter

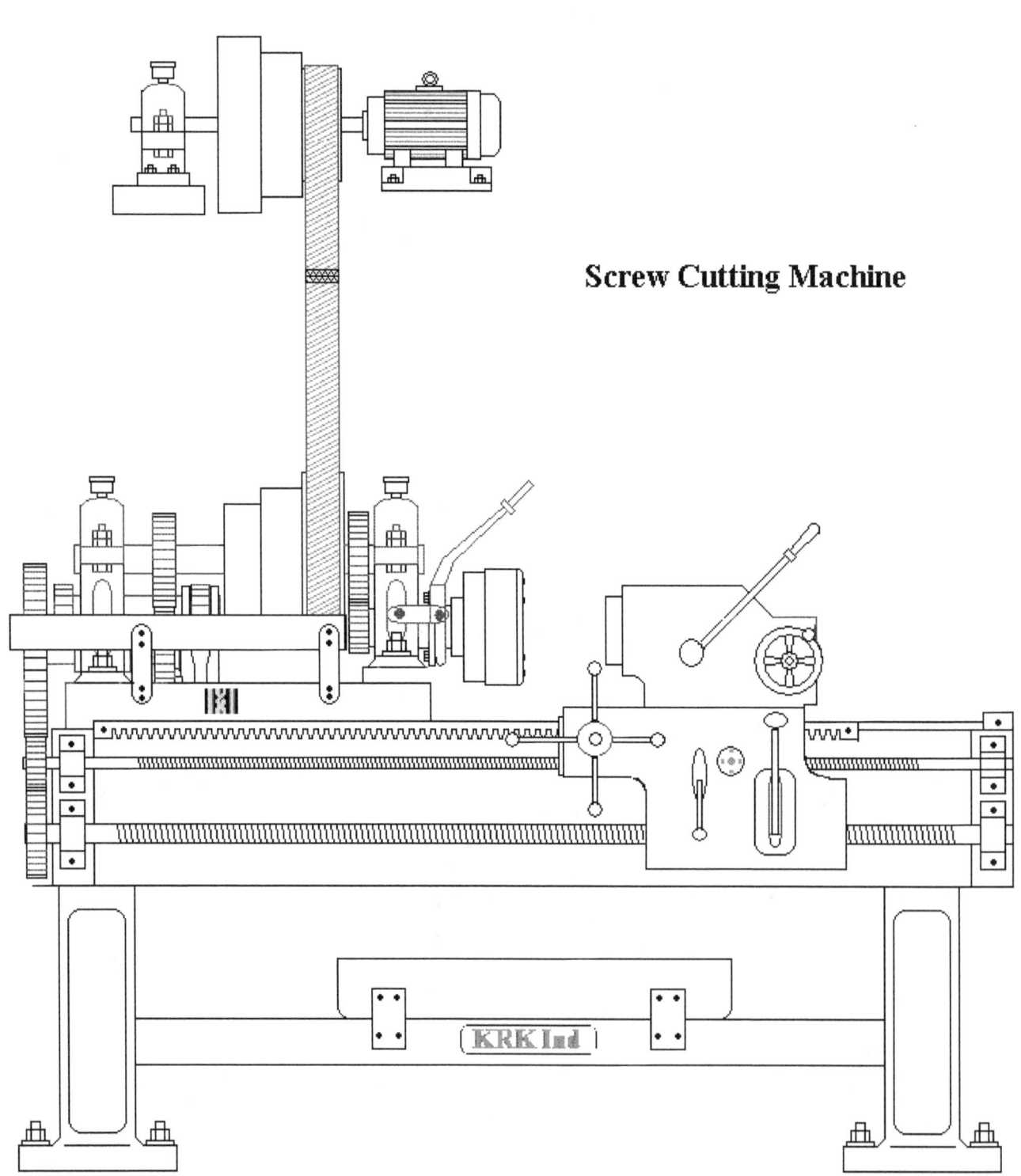

Screw Cutting Machine

KRK Ind

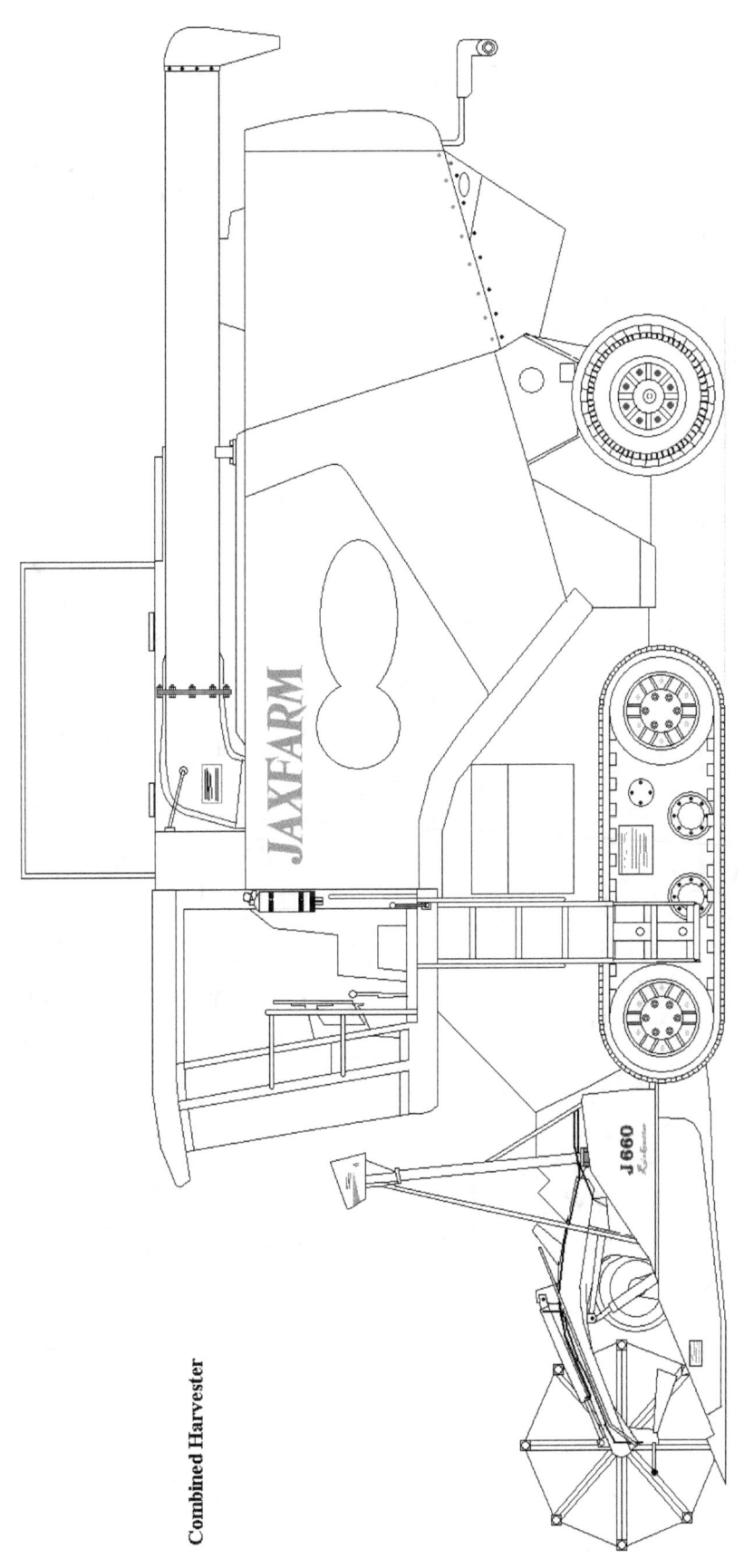

Combined Harvester

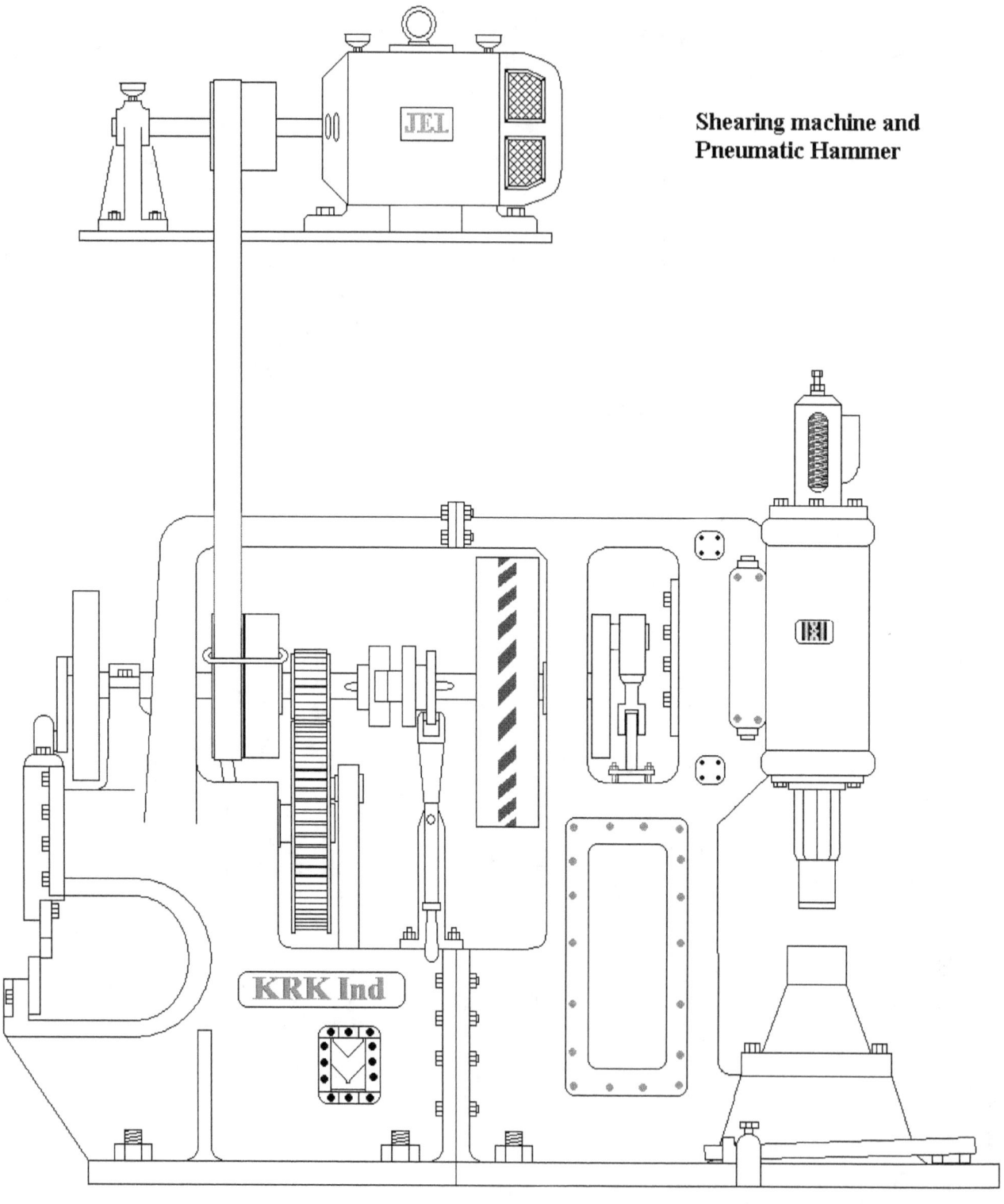

**Shearing machine and
Pneumatic Hammer**

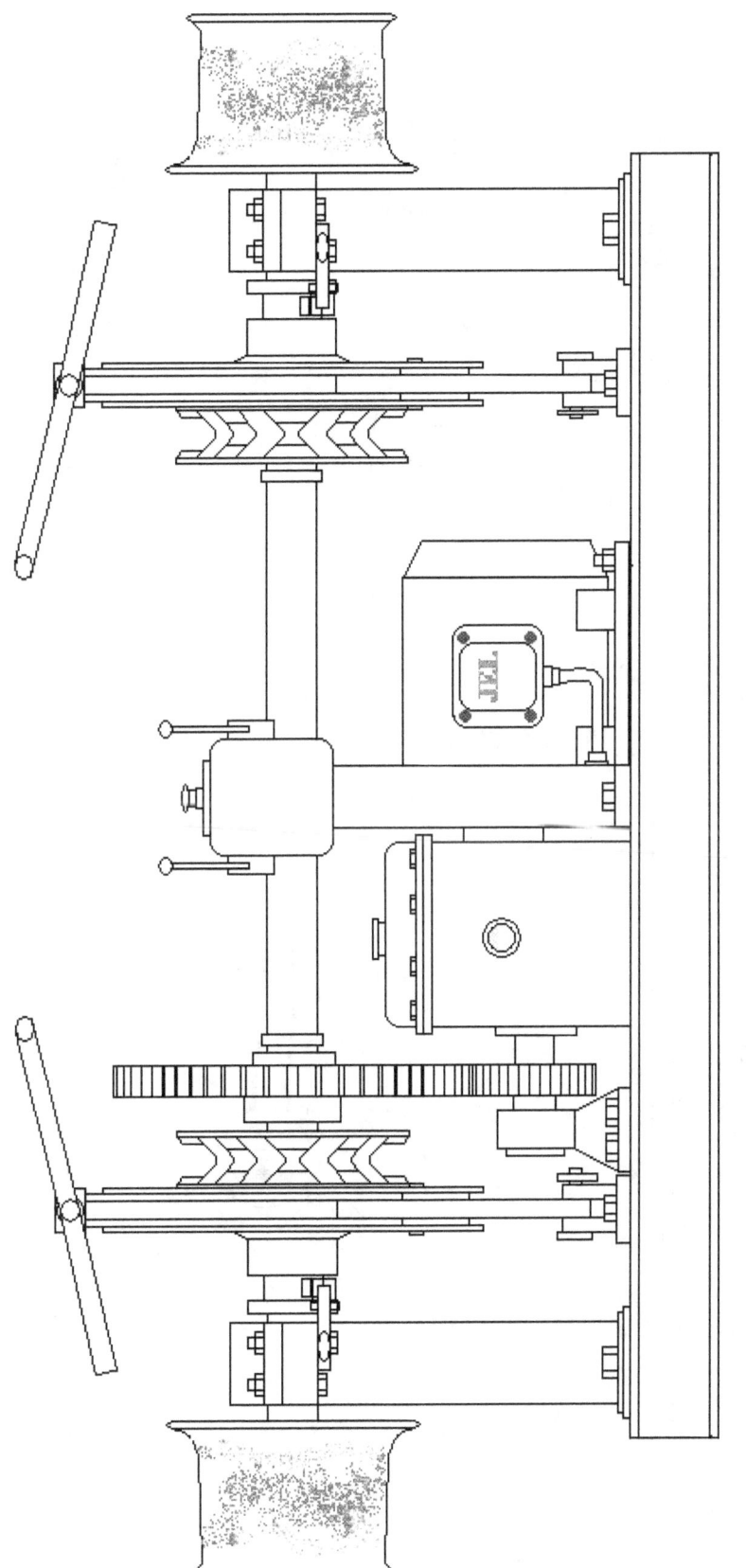

Ship's anchor & rope Windlass

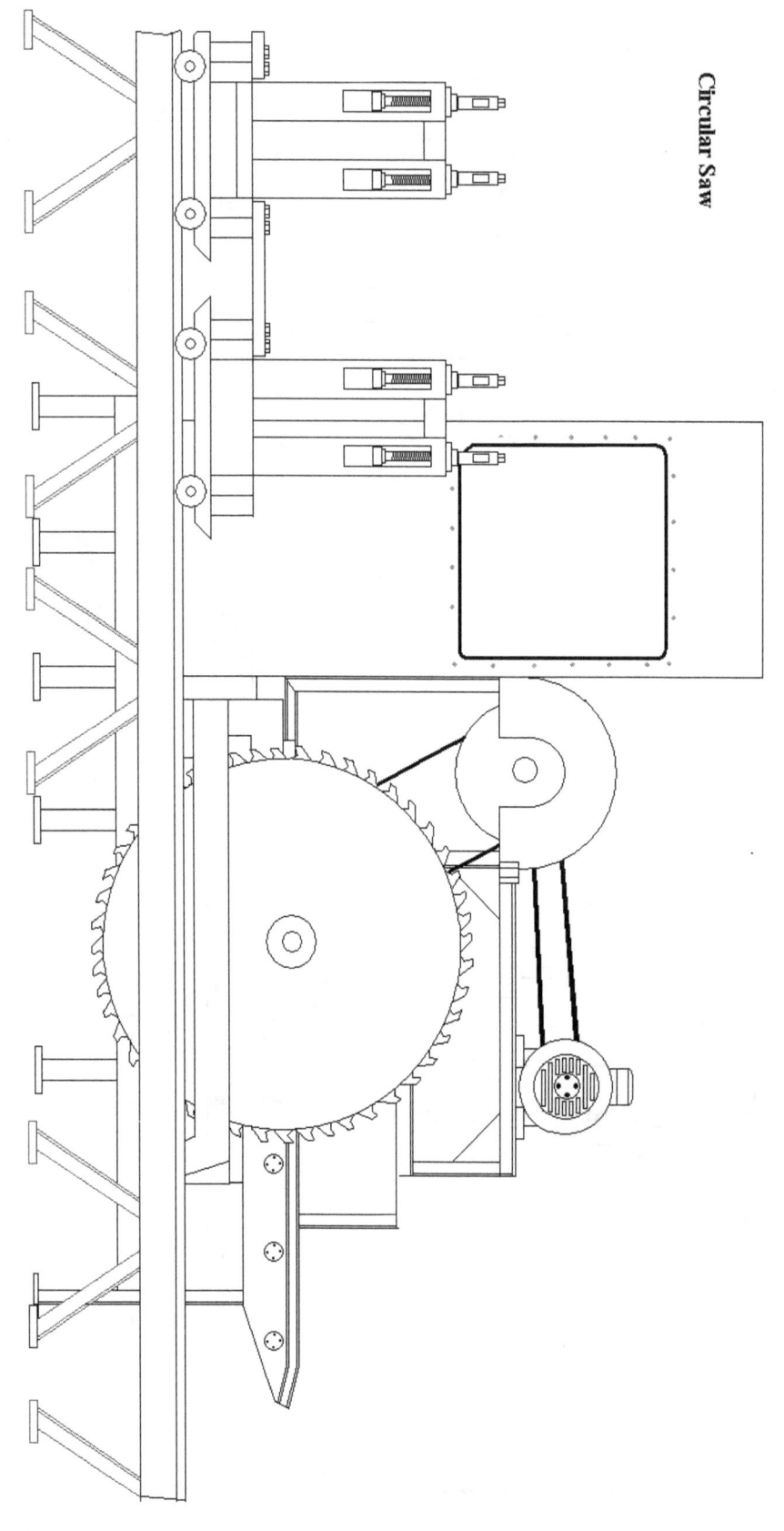

Circular Saw

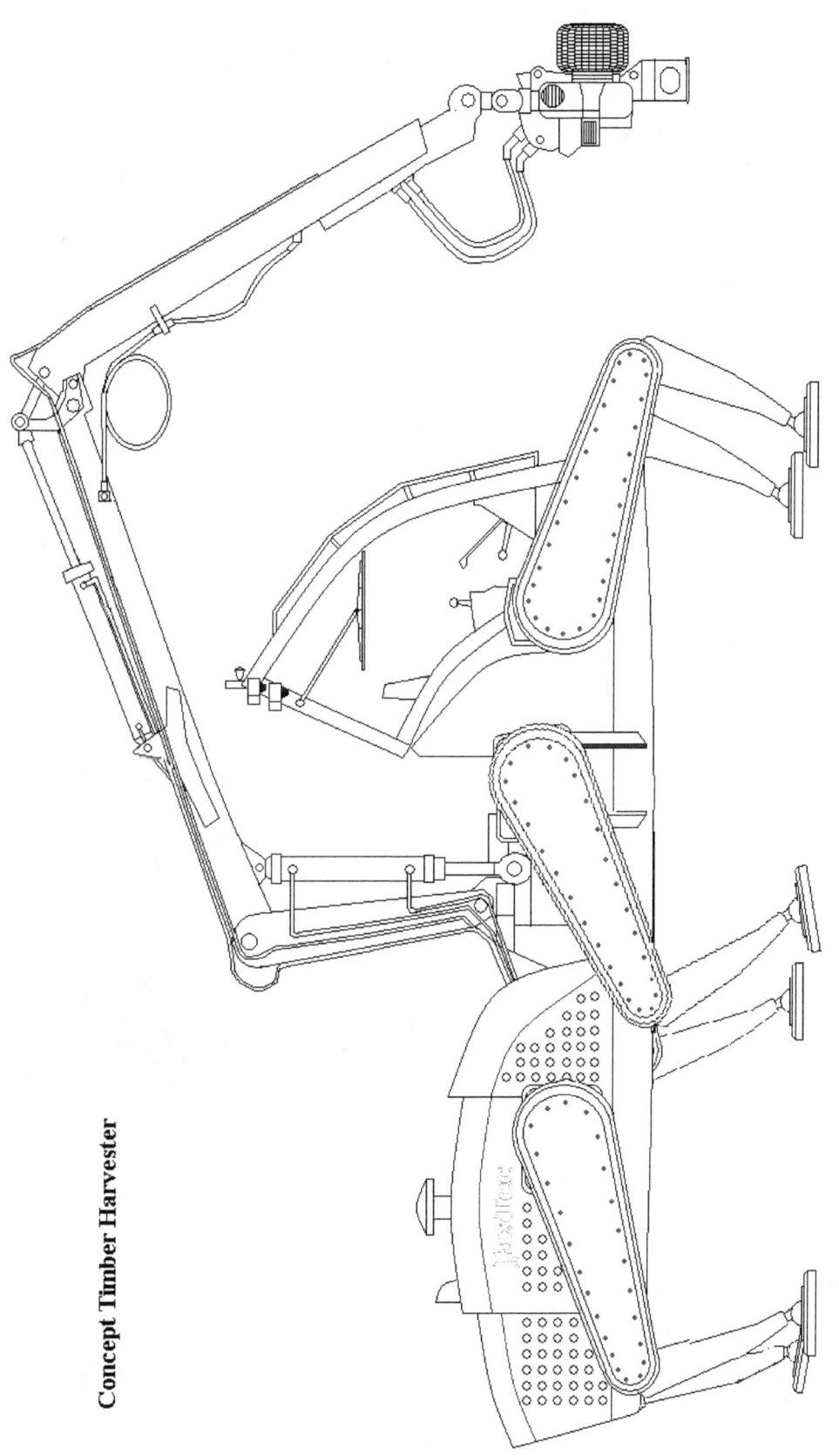

Concept Timber Harvester